Fishing for Fairness

Poverty, Morality and Marine Resource Regulation in the Philippines

Asia-Pacific Environment Monograph 7

Fishing for Fairness

Poverty, Morality and Marine Resource Regulation in the Philippines

Michael Fabinyi

Australian
National
University

E PRESS

ANU
E PRESS

Published by ANU E Press
The Australian National University
Canberra ACT 0200, Australia
Email: anuepress@anu.edu.au
This title is also available online at: http://epress.anu.edu.au/

National Library of Australia Cataloguing-in-Publication entry

Author: Fabinyi, Michael.

Title: Fishing for fairness [electronic resource] : poverty, morality and marine resource
           regulation in the Philippines / Michael Fabinyi.

ISBN: 9781921862656 (pbk.) 9781921862663 (ebook)

Notes: Includes bibliographical references and index.

Subjects: Fishers--Philippines--Attitudes.
          Working poor--Philippines--Attitudes.
          Marine resources--Philippines--Management.

Dewey Number: 333.91609599

Cover design and layout by ANU E Press

Cover image: Fishers plying the waters of the Calamianes Islands, Palawan Province, Philippines,
2009.

Contents

List of Tables

List of Maps

List of Figures

List of Plates

Foreword

Until relatively recent times, the coastal regions of insular Southeast Asia have had the elastic ability to absorb a variety of transient fishing populations from other islands who then within a generation or so assimilate to a new, more fluid ethnic identity. As new technologies have enabled fishing populations to expand ever further in search of lucrative fishing grounds, movement to new settlement areas and the amount of inter-island traffic and fish exports have increased apace. These processes are especially common in the Philippines, and Palawan in the west-central part of the archipelago has attracted many migrant groups in recent decades. As Michael Fabinyi makes clear in his book, *Fishing for Fairness,* the political and economic construction of Palawan as the 'final frontier' and an ecological oasis in the country's overall environmentally damaged set of natural resources, exists in an uneasy tension with the provincial government's strategy of mineral extraction and economic development.

In recent decades, maritime anthropology has really become applied environmental anthropology as national and international efforts to improve the management of marine resources have taken a more interdisciplinary direction. One of the most important ethnographic contributions of such research has been the documentation of existing forms of sea tenure and other territorial forms of dividing up human access to marine resources. A second area of ethnographic contribution concerns the documentation of community-based natural resource management and co-management projects wherein coastal fishers, traders, governments, and often non-governmental organisations design, implement and monitor the use of coastal resources. The relatively poor success record in the Philippines of many of these projects is a theme of this book, as Fabinyi makes the case that unless we understand the narratives and meanings different sets of actors attach to political or environmental initiatives, the effective design of conservation projects are likely to fail—especially if effective alternative livelihood projects are not implemented.

This book is a major contribution to environmental anthropology generally and to political anthropology in the Philippines especially. Despite several decades of global efforts to stop environmental degradation and biodiversity loss in the country, most maritime conservation schemes have failed or, at best, only partially succeeded. Much anthropological literature has corrected Garrett Hardin's critique of open access resources by noting that he confused open access resources with restricted access resources and omitted any consideration of cultural norms and social institutions that might constrain over-exploitation (see McCay and Acheson 1987; Anderson and Simmons 1993; Dyer and McGoodwin 1994). In contrast, Fabinyi's post-structural ecology approach fits into a more recent set of approaches that seek to understand the rhetorics and practices that surround debates about environmental projects from a variety of different perspectives. Here the emphasis is on how such rhetorics and discursive conflicts serve different interest groups and shape institutions, environmental actions, and ways of life (see Greenough and Tsing 2003).

Drawing on cultural insights from a range of scholars specialising in the Philippines, the author highlights how the local conceptions of small-scale fishers of Coron in the Calamianes Islands are expressed through a rhetoric or discourse that equates their fishing practices with legality, morality and self-assessment of their neutral impact on the environment. This politicised rhetoric is formed in a discursive contestation with an opposed image of wealthy fishers associated with illegal fishing practices, immorality and relative impunity from government regulation—in part because they can avoid arrest by paying bribes. Small-scale fishers' construction of wealthy fishers in this way serves as an active campaign of resistance and an effort to prevent government regulation of their extractive practices. Drawing on the 'right to survive' ethics that often color the discourse between the poor and the wealthy in the Philippines, Fabinyi calls this rhetoric 'the discourse of the poor moral fisher'. In a stunning revocation of provincial government efforts to institute a closed season for the live fish trade in the Calamianes, Fabinyi carefully shows how such rhetorics and alliances between small-scale fishers, live fish traders and municipal politicians effectively repelled the provincial regulations. The pro-conservation, provincial politicians were seen by local-level fishers as enveloped within a wider framework that presents almost all actions of fisheries governance as anti-poor and corrupt.

By illustrating his ethnography with a careful review of the work of such scholars as Fanella Cannell, Rey Ileto, and Vicente Rafael concerning how power is conceptualised by marginalised groups in the Philippines, this book contributes rich material to understanding the creative micro-politics and practices surrounding environmental discourses. Rather than examine how cultural norms and sanctions exist to protect certain environmental measures, he instead shows how a culturally constructed image of the larger political economy itself constrains the effective implementation of conservation measures. Fabinyi firmly locates inshore fishers faced with the introduction of marine protected areas within a 'rights-based discourse' that absolves them of responsibility for certain quasi-legal forms of fishing undertaken by young men. He clearly illustrates how negotiable and culturally constructed forms of legality must be understood as embedded in a larger set of perceptions and debates that historically developed to govern relations between small capture fishers and wealthier fishers, traders, and local government officials who are expected to protect their constituents' rights to a livelihood.

Fabinyi makes three main ethnographic points that have public policy implications for the design of coastal resource management policies. First, concepts such as 'legal' and 'illegal' are culturally imbued and flexibly negotiated by very specific resource users according to different sets of resources. Second, marine protected areas are often not able to serve the interests of multiple stakeholders, in contrast to what has sometimes been claimed. Finally, the 'realpolitik' of marine resource management occurs within the discourse of 'fairness' and social justice and itself shapes the outcomes in ways that may undermine the goals of conservation and simply reproduce structures of political patronage. The discourse of morality and fairness toward the poor is a political platform poor people appeal to whenever they are faced with hardship. While patron-client relations are common in Southeast Asia, Fabinyi attributes the emphasis on morality in such appeals to the possible entrenched influence of Roman Catholicism in the Philippines.

This book is designed to be of use to policymakers as well as scholars, including practitioners involved in the implementation of marine resource conservation programs. As such, it is mercifully free of jargon or heavy theory, while providing a carefully nuanced and documented set of insights into the coastal politics of contemporary resource conflicts. By showing how the grievances of fishers are constrained and embedded in larger cultural idioms, Fabinyi's book is a persuasive call for more careful ethnographic documentation both before and during the implementation of co-management projects. It is also a warning of the consequences of not doing so. Given the extreme challenges facing both contemporary fishers and environmental managers in coastal areas of the Philippines, this book is a sensitive but also sobering look at the choices ahead.

Susan Russell
Northern Illinois University
September 2011

References

Anderson, T.L. and R.T. Simmons (eds), 1993. *The Political Economy of Customs and Culture: Informal Solutions to the Commons Problem*. Lanham (MD): Rowman and Littlefield.

Dyer, C.L. and J.R. McGoodwin (eds), 1994. *Folk Management in the World's Fisheries: Lessons for Modern Fisheries Management*. Niwot: University of Colorado Press.

Greenough, P. and A. Tsing (eds), 2003. *Nature in the Global South: Environmental Projects in South and Southeast Asia*. Durham (NC): Duke University Press.

McCay, B.J. and J.M. Acheson (eds), 1987. *The Question of the Commons: the Culture and Ecology of Communal Resources*. Tucson: University of Arizona Press.

Acknowledgements

This book was derived from my doctoral thesis research undertaken at The Australian National University (ANU) from 2005–09. I felt strongly supported by my supervisory panel throughout the course of my research. Colin Filer, the chair of my panel, gave generously with thought-provoking feedback that challenged me to keep the bigger picture and argument at the forefront while writing. Deirdre McKay, a co-supervisor, provided prompt and astute feedback and advice, and was obliging in sharing her insights on the cultural life of the Philippines. Lesley Potter, a co-supervisor, provided very detailed and helpful supervision in the writing up phase and was invaluable in keeping my writing grounded and well structured. Simon Foale, an advisor, gave constructive advice throughout, especially with biological and management aspects of the study. I would also like to acknowledge early support from two anthropologists from the University of Melbourne, Monica Minnegal and in particular Martha Macintyre, who stimulated my interest in anthropology.

Various institutions in Australia and the Philippines have provided support during the course of my research and since completing my thesis. The Resource Management in Asia-Pacific (RMAP) Program at the ANU has offered a friendly and helpful environment to write and research the thesis, and to have ongoing connection as a Visiting Scholar since my conferral in 2009—thanks in particular to Alison Francis, Helen Glazebrook, Andrew Walker, and to Mary Walta for her work in editing the book. Everyday financial support for the Ph.D. was provided by an Australian Postgraduate Award from the Australian Government, and an ANU Supplementary Scholarship. Jenny Sheehan at the ANU College of Asia and the Pacific Cartography Service kindly provided some of the artwork, and I would like to thank Al Linsangan for the use of his photograph that appears as Plate 2-7.[1] I am grateful to the School of Social Sciences at the University of Queensland for hosting me as a Visiting Scholar during my final write-up— in particular to David Trigger and Geoffrey Lawrence. In Manila, the Institute of Philippine Culture at Ateneo de Manila University—especially Cecille Bartoleme—was helpful both in providing an academic base and in navigating my way through various bureaucratic and visa requirements. Stuart Green was helpful in facilitating introductions when I first arrived in Manila, as Arlene Sampang has been in Coron. Since completing my Ph.D., I have revised the manuscript for publication while based at the Australian Research Council Centre of Excellence for Coral Reef Studies at James Cook University, Queensland.

---

1 All other photographs that appear in this manuscript were taken by the author during 2005–10.

For providing helpful comments and advice on sections of the manuscript at different stages, I thank Filomeno Aguilar, Dante Dalabajan, Wolfram Dressler, Helen Fabinyi, Benedict Kerkvliet, Magne Knudsen and Davey Wallace. Discussions with Hannah Bulloch, Amanda Cahill, Aileen Paguntalan-Mijares and Shio Segi were most useful in helping me approach my research. Thank you to the two examiners of the original Ph.D. thesis, Tania Murray Li and Susan Russell, and to the external book reviewer, whose constructive comments have very much improved the thesis.

Since 2007, a number of articles based on doctoral research have been published in academic journals: information from Chapter 5 was published in *Marine Policy* (Fabinyi 2008); sections of Chapter 6 were published in *Philippine Studies* (Fabinyi 2007) and *Coastal Management* (Fabinyi et al. 2010); and an earlier version of Chapter 7 was published in *Human Organization* (Fabinyi 2009).

Unfortunately, because of the need for confidentiality of informants, I am unable to name those who contributed most to the research and the writing of this book—the residents of Coron, 'Esperanza' and others in the Calamianes with whom I worked during my fieldwork. I will not name any individuals from this area because of the sensitivities associated with some of my research, but I am extremely grateful to everyone who either assisted in the research or offered support in other ways. I give particular thanks to the municipal council of Coron and to the council of the *barangay*[2] where I lived, for their hospitality and allowing me to conduct my research. Also, the employees in a number of environmental organisations were most obliging in facilitating aspects of the research. In Esperanza, my host family was extremely helpful in introducing me to the people and place, as well as providing generous family hospitality. They made my fieldwork far more enjoyable and secure with their support. Informants, that were my friends in Coron town, such as many of the tourism operators, were also very generous with their time. I would particularly like to thank my Tagalog tutor, who has always been ready to help.

Other friends and family provided support in other ways: thanks to Nick who came to the Philippines for a visit during fieldwork; Nathaniel and Will for hosting me on my return visits to Canberra; Sarinda for offering personal and academic encouragement throughout; and most of all to my family: my brother and sister, and especially my parents.

---

2   A *barangay* is the smallest political unit in the Philippines.

References

Fabinyi, M., 2007. 'Illegal Fishing and Masculinity in the Philippines: A Look at the Calamianes Islands in Palawan.' *Philippine Studies* 55: 509–529.

———, 2008. 'Dive Tourism, Fishing and Marine Protected Areas in the Calamianes Islands, Philippines.' *Marine Policy* 32(6): 898–904.

———, 2009. 'The Politics of Patronage and Live Reef Fish Trade Regulation in Palawan, Philippines.' *Human Organization* 68: 258–268.

Fabinyi, M., M. Knudsen and S. Segi, 2010. 'Social Complexity, Ethnography and Coastal Resource Management in the Philippines.' *Coastal Management* 38: 617–632.

Selected Tagalog Glossary

Tagalog	English
amihan	northeast monsoon, also called the dry season, from October to early May
awa	pity
bale	advance payment
barangay	smallest political unit in the Philippines
basnig	bag or lift-net boats
bayanihan	cooperative work day
cascasan	the hook-and-line method used in fresh grouper fishing
dilis	anchovies
habagat	southwest monsoon, also called the wet season, from late May to September
hiya	shame, shyness or embarrassment
hulbot-hulbot	Danish seine
kawawa	pitiful
kusinero	cook
kuya	older brother
largo viaje	fishing method using heavy lines with multiple hooks
malaking tulong	big help
makatao	concern for humanity
mekanista	engine operator
muro-ami	illegal and destructive fishing technique
pakikisama	togetherness, an ability to get along with people
pangulong	baby purse seine
po	Tagalog grammatical particle indicating respect
sangguniang bayan	municipal council
sari-sari store	general store
sitio	enclave within a barangay
suki	a form of personalised economic relationship
ulam	main dish accompanying rice
utang	debt
utang na loob	debt of the heart/inside
videoke	karaoke

Abbreviations

BFAR	Bureau of Fisheries and Aquatic Resources
CI	Conservation International
CLOA	Calamianes Live-fish Operators Association
ELAC	Environmental Legal Assistance Centre
DENR	Department of Environment and Natural Resources
FISH	Fisheries Improved for Sustainable Harvest
ICDP	Integrated Conservation and Development Project
ICM	Integrated Coastal Management
IPC	Institute of Philippine Culture
LGC	Local Government Code
LGU	Local Government Unit
MEA	Millennium Ecosystem Assessment
MPA	Marine Protected Area
PCSD	Palawan Council for Sustainable Development
SEMP-NP	Sustainable Environment Management Project in Northern Palawan
SEP	Strategic Environmental Plan
UNESCO	United Nations Educational, Scientific and Cultural Organisation
WCED	World Commission on Environment and Development
WWF	World Wide Fund for Nature, now known as 'WWF, the Conservation Organisation' in the Philippines

Currency Conversion Rates

Year	1US$ equivalent in ₱ (PHP)
1993	27.4
1994	24.2
1995	26.2
1996	26.3
1997	40.0
1998	38.8
1999	40.2
2000	49.9
2001	51.5
2002	53.2
2003	55.2
2004	56.1
2005	52.9
2006	48.8
2007	41.1
2008	47.2

1. Introduction: Fishing for Fairness

This book is an analysis of how local coastal communities in the Calamianes Islands in the Philippines understand the relationship between power, wealth and the environment, and how this understanding has contributed to the current situation of marine resource management. Unlike perspectives that have sought to establish objective measures of this relationship, I am interested in how it is subjectively understood and represented, by examining how local discourse has shaped a process of contestation over marine resources. Such management contestations are a characteristic feature of the 'resource frontier' in the Calamianes Islands, where fishing, conservation and tourism enact competing visions of how to engage with the bounty of marine resources.

Fishers in the Calamianes Islands with whom I have worked represent their fishing traditions as possessing two key features: their fishing methods are harmless to the environment and their use of low technological gear is closely tied to their poverty. Because of this, their practices are imbued with a sense of morality. In contrast, the activities of 'immoral' illegal fishers and outsiders are perceived as being responsible for all environmental degradation. From this perspective, it follows that any regulations introduced by government to reduce environmental problems should address those who cause the problems (the illegal fishers) and those who can afford to pay for their amelioration (the illegal fishers and the tourism industry).

This local fisher discourse was expressed, with varying emphases, in a range of contexts concerning marine resource regulation in the Calamianes. Two notable cases in point occurred during the debates on reforming the regulations governing the lucrative live fish trade, and implementing a series of marine protected areas. By adopting this discourse in these debates, fishers contributed greatly to the decisions reached, namely: the overturning of the live fish trade regulations and changes in the proposed implementation for the marine protected areas. Understanding the nature and effects of what I call the discourse of the 'poor moral fisher'[1] became the primary focus of my research.

While it is possible to argue that this was simply a strategic political ruse by the fishers to avoid any regulation that would affect their fishing practices, here I explore these stated beliefs in greater depth—unraveling their relationship to cultural and social institutions. I found that the discourse of the poor moral fisher is fundamentally embedded within strongly held ideas and practices in

---

1   There are two points to clarify about my use of the term, 'poor moral fisher'. Firstly, by using this term I mean that the discourse emphasises that fishers are poor and moral; not that they have poor morals. Secondly, as will be demonstrated throughout the book, the term is not meant to imply a sense of absoluteness. I use the term more for a sense of narrative clarity, and instead emphasise the varied ways in which it is expressed.

the Philippines. I show how it is intertwined within a 'basic rights discourse' (Kerkvliet 1990: 242–73; Cannell 1999: 231–4) that is commonly expressed in the Philippines, and that it is reflected in particular ideas about reciprocity, social obligations and morality. These local concepts go a great way to explaining the context of the discourse of the poor moral fisher, and the means by which it achieves such resonance.

Such an emphasis on culture and morality draws on two general insights about the value of such topics in political ecology. Firstly, as Brosius has argued, '[w]e have been so fixed on local social movements, transnational NGOs, and globalizing processes that we seem to have forgotten about the need to understand how national political cultures might mediate between these' (Brosius 1999: 285). In the Philippines, the national political culture that I emphasise is that of the 'basic rights discourse', which stresses social obligations and moral relationships between rich and poor. This focus on morality as it is understood in the Philippines engages with Bryant's (2000) broader argument about the relevance of morality in debates over the environment: 'Research in political ecology has paid inadequate attention to the multifaceted cultural politics surrounding discourses of environmental conservation in the developing world. Specifically, it has tended to neglect the rich politicized moral geographies integral to conservation debates' (ibid.: 673, see also Bryant 2005).

In a regional academic context, this book will show how anthropology can make a distinctively cultural contribution to debates about environmental politics in the Philippines. I will do this by using the insights about local conceptions of power, reciprocal relationships and morality in the Philippines to inform my discussion of the politics of environmental management. I shall demonstrate how what may on the surface appear to be straightforward responses to environmental regulations are also about an attempt to modify the social and political relationships that fishers maintain with various other actors: an attempt to 'fish for fairness'. I argue that the responses of fishers to environmental regulations should be understood firmly within the context of these relationships.

By adopting a theoretical framework of a particular form of political ecology that emphasises the role of discourse and culture in environmental politics, the book also aims to extend ways of understanding the poverty-environment relationship. Without ignoring material practices and reality, I show that perceptions of the poverty-environment relationship make an important contribution to material outcomes. Like Brosius, I argue that discourse plays a significant role: 'discourse matters ... environmental discourses are manifestly constitutive of reality (or, rather of a multiplicity of realities)' (Brosius 1999: 278). Discourses about the poverty-environment relationship, I argue, can be seen not only in the words of fishers, but in the broader cultural setting of the Philippines where ideas about morality and reciprocity are elaborated on and acted out in many contexts. As Hall (1997: 44) has pointed out, '[i]t is important to note that the concept of discourse in this usage [that of Foucault, and of mine in this book] is not purely a "linguistic" concept. It is about language and practice ... his definition of discourse is much broader than language'.

In this introductory chapter, I begin by reviewing the aspects of academic literature and global debate that focus on understanding the relationship between poverty and the environment. Further, I situate my research by introducing the features of post-structural political ecology that have influenced my theoretical perspective. After briefly contextualising my research within other work in the Philippines and within the broader concerns of anthropology, I continue with a discussion of the research methods I used, and then outline the rest of the book.

Poverty-Environment Relationship: Conceptualisations

Sustainable Development

Genealogies of debates about the poverty-environment relationship (for example Gray and Moseley 2005) have often started with a reference to the 'sustainable development' discourse, epitomised in the 1987 report by the World Commission on Environment and Development (WCED), more commonly known as the Brundtland Report. In a frequently quoted passage, the Brundtland Report declared that: 'Many parts of the world are caught in a vicious downwards spiral: poor people are forced to overuse environmental resources to survive from day to day, and their impoverishment of their environment further impoverishes them, making their survival ever more difficult and uncertain' (WCED 1987: 27).

This view of sustainable development attempted to move beyond earlier perspectives that viewed environmental concerns and economic development as fundamentally opposed. It aimed to address both the interests of observers in richer countries concerned about environmental degradation in the developing world and the interests of developing country governments more focused on economic growth. The sustainable development perspective was supported through its appearance in various international conventions and conferences, and by the early 1990s had become the primary view underlying massive funding for integrated conservation and development projects (ICDPs) by a range of international organisations (Wells et al. 2004: 401). Adopting the view of poverty and environmental degradation as mutually reinforcing, ICDPs frequently targeted the activities of poor people, assuming that they were the central threat to the environment (ibid.: 406).

A second approach to the poverty-environment relationship has aimed at infusing it with an explicitly political perspective—political ecology. This approach is similar to the perspective of fishers in the Calamianes. This materialist form of political ecology, which emphasises the structures behind poverty and the impact of wealth on environmental degradation, is one which Brosius (1999: 303) describes as 'representing a fusion of human ecology with political economy'. It tries to explain environmental politics and poverty-environment interactions in terms of empirical factors and causes.

Materialist Political Ecology

While the field of materialist political ecology has continued to rapidly expand and diverge in terms of its thematic interests (see Bryant 1992; Peet and Watts 1996; Robbins 2004; Paulson and Gezon 2005), one of the core themes has remained its attention to the relationship between poverty and environmental degradation. Robbins characterised the argument formed from the ways these political ecologists have examined this relationship as the 'degradation and marginalization thesis' (Robbins 2004: 14). As Robbins described it, the essential feature of this thesis was that '[l]and degradation, long blamed on marginal people, is put in its larger political and economic context' (ibid.). This had the effect of 'shifting the blame' from poor people to the things that were making them poor. Blaikie and Brookfield (1987), for example, argued that peoples' political and social marginalisation correlated with their use of ecologically marginal land, which because of increasing human demands exacerbated environmental degradation. Earlier work by Blaikie (1985: 147) similarly emphasised how soil erosion was inevitably linked with issues underlying poverty and underdevelopment: 'small-scale land-users often directly cause soil erosion, because they are forced to do so by social relationships involving surplus extraction'. From these works, the forces and social relations behind poverty are seen as the 'real' cause of environmental degradation.

Other works in this area of political ecology have continued to focus attention on the theme of poverty-environment interactions, but have taken somewhat different approaches. Gray and Moseley (2005: 19), for example, argue that '[w]hile poverty may be an important driving force of environmental degradation … wealth and economic development are more likely culprits'. This was typified by the situation at Moseley's fieldsite in Mali where it was found that wealthier cotton farmers had a greater negative environmental impact than the poorer ones (Moseley 2005).

The 'downward spiral' model and the materialist political ecology model represent attempts to understand the poverty-environment relationship from two different perspectives. However, as recent models of the relationship such as the ecosystem approach (MEA 2003) and that taken by the United Nations Poverty-Environment Initiative (Comim 2008) suggest, the particular features of the relationship can vary dramatically depending on location, time and scale. This has led to the conclusion expressed in some policy studies that indicators and models must be grounded and developed with local participation (Reed and Tharakan 2004; WVA 2006). As Comim (2008: 21) argues: 'good poverty and environment indicators need to be integrated and anchored on local values and decision-making processes'. Building on this notion that shifts the focus of study away from some of the empirically measurable features of the poverty-environment relationship, the purpose of this book is to focus instead on the subjective ways in which it is understood, represented and contested among local actors.

Situating the Book

Importance of Discourse and Local Values: Post-Structural Political Ecology

In contrast to previous studies, which have aimed to empirically measure or objectively analyse the poverty-environment relationship, here I consider local perceptions and representations of such a relationship. Ethnography, with its strengths in examining local processes and politics, is particularly well suited to this approach. Its value is that it shows how environmental problems have come to be defined and contested at a local scale. Understanding the discourse of local residents about environmental issues—through an in-depth examination of its cultural and political contexts—goes some way to elucidating the manner in which this discourse has come to shape and influence the marine resource management process in the Calamianes. By showing how the political grievances of fishers are embedded in particular cultural perspectives, I explain the shortcomings of current resource management policies.

This approach draws on the work of Brosius (2001: 151), who emphasises that 'environmental discourses configure (or are in turn configured by) emerging forms of political agency'. For Brosius, environmental campaigns:

> ... are not merely the sum total of a series of points of contestation among a range of actors with a diversity of fixed perspectives.... In the process, certain actors have been marginalized, while others have been

privileged. If anthropologists are concerned with understanding the processes by which emerging forms of political agency are constituted and configured, it defeats our purpose to regard these debates in terms of mere polyvocality. In fact, certain voices are able to edge others out, certain voices may be co-opted, certain voices may be dismissed as disruptive, and certain forces may be taken as irrelevant. How does the process of forcing open spaces for newly emerging political agents occur? How or why do such spaces close on others? (ibid.).

Similarly, in his study of discursive conflict among fishermen in Brazil, Robben (1989) has used discourse analysis to show how 'the economy' is enacted, represented and contested in different ways by different groups of fishermen. Here, various discourses about what is or is not 'economic' among more or less powerful groups have important implications for the distribution of benefits in the fishing sector.

By following this approach to understanding the roles of discourse, this book can therefore be viewed as contributing to the broad strand of literature on the politics of environmental use that has been labelled as 'poststructural political ecology' (for example Peet and Watts 1996). In contrast to the materialist form of political ecology, such literature is characterised by the analysis of both material and discursive contestation, arguing that practical struggles are always simultaneously struggles for 'truth'—struggles that happen in imagination and representation at the same time as they are conducted in the material world (ibid.: 37). This form of political ecology focuses on 'how competing claims to resources are articulated through cultural idioms in the charged contests of local politics' (Moore 1996: 126; see also Alejo 2000; Walley 2004; West 2005). I aim to show how local conceptions of the poverty-environment relationship are manifested in debates about regulation, and are embedded in worldviews and social relations that are specific to the Philippines.

Much could be learned from an integrated study of quantitative and qualitative aspects of the poverty-environment relationship, but my goal here is a preliminary step towards this end. I map out the cultural dimensions in a qualitative way, using some ecological and economic data as a setting to the primary focus on social and political contestation. Similarly, although it would have been useful to include more detailed analyses of other perspectives on environmental degradation, such as from conservationists, local governments or tourism operators, I focus on these attitudes more as they relate to the specific arguments set forth by local fishers. My primary aim is to show how resident fishers understand the various links between power, wealth, and environmental degradation, and how these understandings inform and help shape the outcomes of local debates about regulation of marine resources.

1. Introduction: Fishing for Fairness

Approaches in the Philippines

Numerous studies have examined issues relating to the poverty-environment relationship and environmental politics in the Philippines. The large interest in this topic is a function of factors that include the: high levels of biodiversity (Carpenter and Springer 2005); high rate of poverty; and a significant number of donor-funded biodiversity conservation and poverty alleviation projects in operation. The Philippines is therefore an excellent site to examine poverty-environment interactions.

One body of literature has analysed the impacts, successes and failures of developments in coastal resource management in the Philippines (for example Courtney and White 2000; Pollnac et al. 2001a; Christie et al. 2005, 2009; White et al. 2005).[2] Aspects of coastal resource management that have been analysed include marine protected areas, legislation to promote decentralised management, and specific project experiences. Several of these researchers participated in a broad study of 'integrated coastal management' (ICM) in the Philippines and Indonesia and published their findings in a special issue of Ocean and Coastal Management in 2005. Their aim was to evaluate the sustainability of this form of management in achieving a range of goals. Overall, they found that ICM was rarely self-sustaining and that environmental conditions continued to worsen. They attributed this failure to the difficulties associated with building new institutions in a developing country context such as the Philippines (Christie et al. 2005).

In contrast to this managerial approach, which has had a more explicit focus on policy, other writers have approached issues relating to environmental use in the Philippines from perspectives informed by a more political perspective. Many of the anthropological discussions of environmental politics in the Philippines have come from scholars working on Palawan, because of its unique environmental and historical features. Here I briefly point out some of the main issues they raise and how my research sits in relation to their findings (as further detailed in Chapter 2).

Themes in this literature include the: analysis of conservation as a form of governmentality (Bryant 2002; Seki 2009); strategies used by environmental NGOs (Bryant 2005; Austin and Eder 2007; Novellino and Dressler 2010); impacts of conservation projects (Eder 2005; Dressler 2009); and how environmentalism in Palawan and elsewhere in the Philippines is marked by a simultaneous concern for social justice (Austin 2003; Bryant 2005). Eder's more recently published

---

2   Coastal resource management initiatives in Palawan and elsewhere in the Philippines are reviewed in more detail in Chapters Two and Five.

study of coastal livelihoods and resource management projects in San Vicente Municipality, on mainland Palawan (Eder, 2008) therefore offers a good contrast to the situation in the Calamianes.

This literature has provided detailed and rich accounts of the relationships between local communities and organisations seeking to implement conservation and management programs, as well as addressing a range of other themes. These writers have demonstrated, in particular, a concern for and understanding of local social and political processes; considerations mostly absent in the management-focused literature on coastal resource management. In contrast to the management literature, the socio-political researchers place the experiences and livelihoods of local communities at the centre of their analyses. However, while such analyses have focused on local social and political concerns, there has been little attention as to how environmental politics in the Philippines are tied to the cultural norms of reciprocity; a topic this book aims to redress.

Fishing for Fairness

The purpose of this research is to build knowledge about local processes through detailing local understandings, attitudes and discourse concerning poverty and the environment. It is my intention to show how the social and political behaviour of local residents is deeply rooted in cultural norms of reciprocity, supported by widely held notions of personal morality and virtue. I will demonstrate how the political grievances of fishers are embedded within particular cultural contexts that determine how environmental politics are played out in the Philippines. A focus on local conceptions of morality among fishers and their relevance to conservation debates is set in a broader context of insights gained by other writers on social life and culture in the Philippines.

In particular, I draw on the work of Cannell (1999), which in turn built on the ideas of Ileto (1979) and Rafael (1988) about conceptualisations of power in the Philippines. Cannell has argued that themes of oppression and power are represented consistently in the Philippines through an idiom of 'pity' (*awa*) (Cannell 1999: 231–4). She shows how different groups of poor people present themselves as 'pitiful' in their relationships with more powerful people in various ways, and that such relationships must be viewed in the context of these ideas. Like Kerkvliet (1990), Blanc-Szanton (1972) and others, Cannell documented the existence of a strong ethic of fairness and justice for the poor.

In the Calamianes, I too noted fishers presenting as 'pitiful' (*kawawa*) specifically for the purposes of establishing reciprocal relations with those more powerful. Here, fishers expect the rich and powerful to recognise their inherent human dignity and treat them with humanity (*makatao*), creating a shared social world. The claims of the poor moral fisher are ultimately about ensuring that the fundamental bases of the relationships that form their social world are respected. In this regard, I will show how fisher folk responses to environmental regulations can be viewed as an attempt to improve the social and political relationships between fishers and people with more resources. By using the ideas of Cannell (1999) and others who have theorised on social relationships in the Philippines, I aim to bring this element of culture to the forefront of environmental politics discussion in the Philippines.

Broader Orientations within Anthropology

Such a concentration on discourse, culture and understanding inevitably leads to broader questions and orientations in anthropology—questions about the relationship between: action and understanding, economy and culture, and humans and the environment. Ecological and environmental anthropologists have long debated these questions. Writers in the culturalist tradition of environmental anthropology have emphasised the ways in which culture, ideas, language and the human imagination serve to shape events and outcomes with regard to environmental issues. This tradition encompasses work such as that of Ingold (1993), a body of literature that has been termed 'symbolic ecology' (Descola and Pálsson 1996), and those who have written with an explicitly post-structural or Foucauldian perspective (Escobar 1999; Bryant 2002, 2005).

While I have a great deal of sympathy for the culturalist perspective, I also take heed of the critique made by Carrier, who cautions that the culturalist orientation of writers such as Ingold (1993) and Escobar (1999) 'tends to ignore the possibility that people's understandings of and actions regarding their natural surroundings may be in a generative, dialectical relationship with each other' (Carrier 2001: 39). In other words, such models can consider people's actions as 'relatively unproblematic reflections' (ibid.: 27) of their views or perceptions of the natural environment. Similarly, typical criticisms of discourse analysis have noted how it can lead to 'out-of-touch' analyses, far removed from the realities of social life (for example Filer 2004: 84).

At the materialist end of the anthropological spectrum, writers emphasise the ways in which social life and culture have been heavily influenced by peoples' material environments. In maritime anthropology, for example, writers have focused on the commonalities among fishing societies, arguing that they are a response to the physical and social characteristics of most fishing communities (for example (Acheson 1981). From this perspective, the social organisation of fishing communities is a response to the distinctive influences of fishing. Pálsson (1991: 38–42), among others, has persuasively demonstrated the tendencies of this 'natural model' of fishing to ignore the many cultural and social differences between fishing societies. He also highlighted the relationship between the 'natural model' and earlier accounts of cultural ecology (Steward 1955).

In this book I do not intend to enter into detailed discussions of causality; whether the superstructure is determined 'in the last instance' by economic and material forces, or whether economic life is a product of the ways that people understand it—writers have for many years debated these questions with little resolution (as reviewed in Wolf 1999: 21–67). Sahlins, for example, highlights the weaknesses of unbalanced visions of both materialism and idealism:

> For materialism, the significance is the direct effect of the objective properties of the happening. This ignores the relative value or meaning given to the happening by the society. For idealism this is simply an effect of its significance. This ignores the burden of 'reality': the forces that have real effects, if always in the terms of some cultural scheme (Sahlins 1985: 154).

Instead, I aim to adopt an approach similar to that of Billig, who has characterised his approach as one which:

> view[s] … symbolic and cultural understandings as resources that are often used strategically to advance or resist "interests" that are themselves culturally constructed in a never-ending chain of mutual and interactive causality. Culture is not a looming, static thing that causes or creates. It is itself always open to negotiation, change, and individual agency (Billig 2000: 782–3).

Writers adopting this perspective, Billig argues, 'view culture within local regional, and global political and economic contexts but appreciate that macro-scale events and effects are always perceived, conceived, and acted upon within culturally constructed meanings' (ibid.: 783). Or as Bertrand Russell (2004 [1946]: 2) suggested perhaps more succinctly: '[t]here is here a reciprocal causation: the circumstances of men's lives do much to determine their philosophy; but, conversely, their philosophy does much to determine their circumstances'. While I recognise the danger of essentialism by adopting an overtly cultural approach in my book, I do not intend to imply that the cultural idioms in this book are unchanging, uncontested, or that they are shared by everyone—simply that they are widely shared and culturally intelligible. The discourse of the poor moral fisher is not some primordial, abstract, cultural system that directs all fishers how to act accordingly; rather, I present it as enacted through aspects of social life and behaviour. As Geertz (1973: 17) stated, 'it is through the flow of behavior—or, more precisely, social action—that cultural forms find articulation' (as cited in Turner 1975: 147). One of the primary contentions of this book is that the conceptions of fishers, as expressed through the discourse of the poor moral fisher, are important to any understanding of the actual material outcomes of marine resource policy in the Calamianes. As Chapters 5 and 7 will reveal in particular, the ways in which this discourse is expressed hold significant implications for the outcomes of various environmental regulatory interventions.

I also seek where possible to avoid some of the tendencies of political ecology towards strict dichotomies. In particular, I avoid what some writers have chosen to characterise as a 'black and white' relationship between powerful, global or foreign conservationists, and weaker, local and/or indigenous groups (for example Brockington 2004). My book does not focus on one single conservation project or the project workers' interactions with local communities. While critical at times about aspects of marine conservation and regulation, I do not intend for the book to become defined as yet another self-righteous Western critique of how conservation projects have marginalised or disempowered local communities. While such critiques can play a valuable role in highlighting social injustice, on their own they frequently do little either to advance the terms of academic debate, or to assist in understanding or dealing with the practical social and ecological problems at hand. Instead, I view the regulatory regimes I studied—regulations to reform the live fish trade, and the creation of marine protected areas—as contexts where local understandings about the relationship between poverty and the environment have played out. Nor do I intend to imply that these regulatory regimes are necessarily representative of some form of globalisation, instead emphasising the local flavour of conservation in the Philippines (Brosius 1999: 285; Austin 2003; see also Filer 2004 for Papua New Guinea).

Another dichotomy often used in political ecology is that between community and the state. Such a dichotomy has frequently led to what Brown and Purcell (2005) call 'the local trap', which 'leads researchers to assume that the key to environmental sustainability, social justice, and democracy (commonly desired outcomes among political ecologists) is devolution of power to local-scale actors and organizations' (ibid.: 608). I aim to avoid such explicit valorisation of local groups. Instead, I emphasise the morally difficult nature of many of the problems related to coastal resource use in the Philippines. By this, I simply mean that the solutions to these problems are never simple or clear-cut, and inevitably involve questions related to values and morality. Such a view attempts to acknowledge the complexity of the issues in a way that both radical critiques of Western development—which can tend to romanticise local groups (for example Escobar 1995)—and ecocentric critiques of community-based conservation (for example Soulé and Terborgh 1999) are unable to do.

Similarly, as Li has argued with reference to the work of Scott (1998), it is important to 'question the spatial optic of Scott's account that posits an "up there", all-seeing state operating as a performed repository of power spread progressively and unproblematically across national terrain, colonizing nonstate spaces and their unruly inhabitants' (Li 2005: 384). While it would have been useful to conduct a truly 'multi-scale' ethnography (Paulson and Gezon 2005) and observe provincial and national political cultures and processes in far greater detail, this was not the focus of my research. Instead, I have aimed to address this issue by drawing particular attention to the ways in which different levels of government are perceived at the local scale (see Chapter 7).

Research Methods

Informants and Techniques

My research was based primarily in two locations: one in Esperanza (a pseudonym), a small peri-urban *sitio* (an enclave within a *barangay*) several kilometres from Coron town, and Coron town itself—the capital of Coron municipality (see Maps 1-1, 1-2). I visited Coron for one month in September 2005 for a 'scoping' trip, and returned in January 2006 to undertake research. Esperanza was a good location for me to work because most of the residents there were fishers, a larger proportion than in other coastal barangays in Coron. Of key interest for my research was local responses to conservation and management regulations, and so Esperanza appeared an opportune place to be because of the MPA (Marine Protected Area) recently implemented there. With the assistance of a local resident with whom I had made contact, I moved into the house of a local fishing family in Esperanza at the beginning of March. Except for a month in July when I returned to Australia, I stayed in Esperanza and Coron until the end of January 2007. I have since returned for trips in March and June–July of 2009, and July 2010.

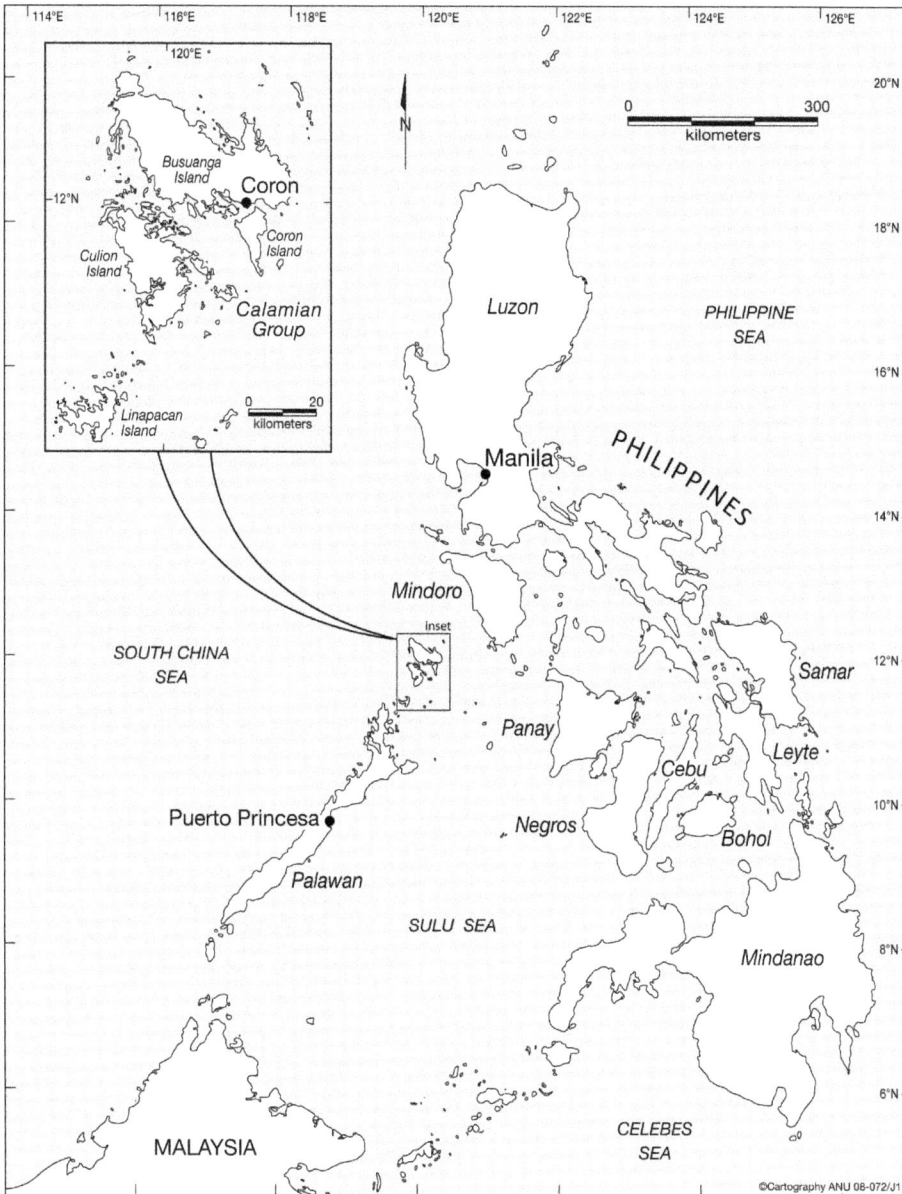

Map 1-1: Map of the Calamianes Islands.

Note: for reasons of confidentiality, Esperanza is not marked on the map. Source: Cartography ANU.

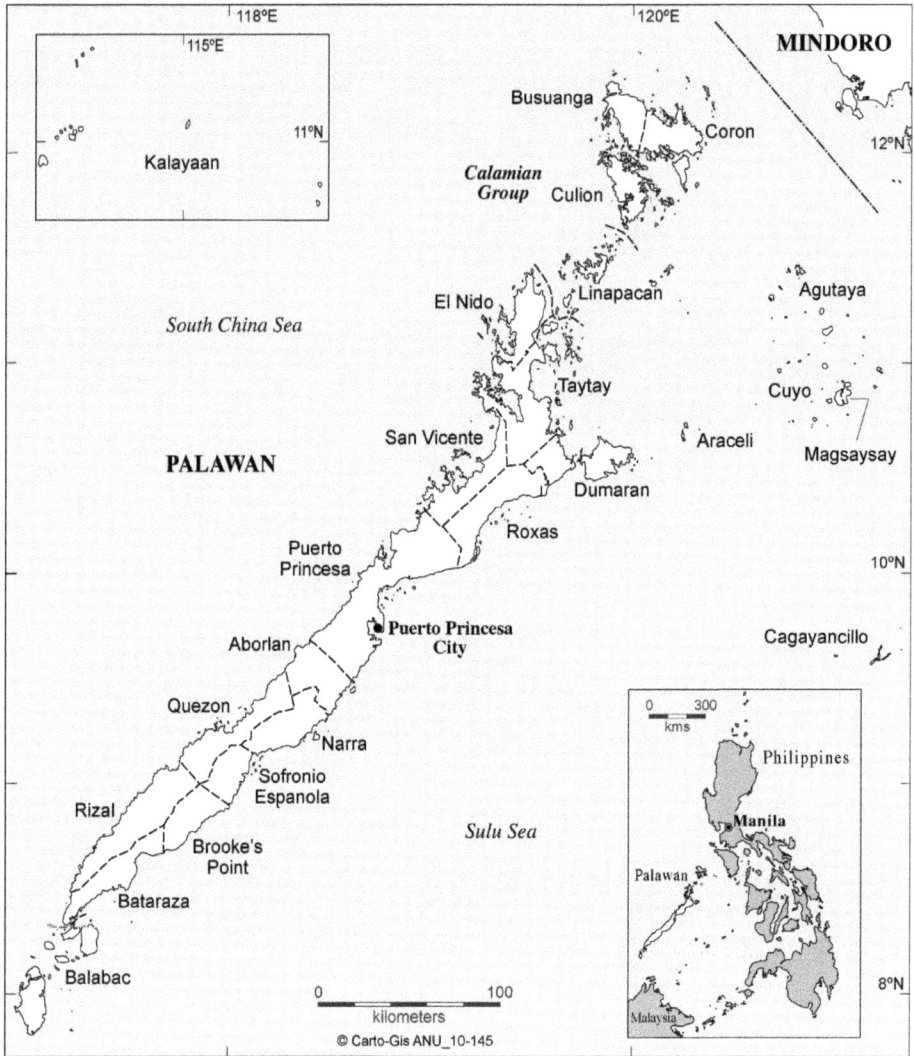

Map 1-2: Map of Palawan Province.

Source: Cartography ANU.

Many scholars working in Palawan have focused on issues to do with indigenous minorities (for example Dressler 2009; Novellino 2007), and this would certainly have been an informative and interesting line of research to adopt in the Calamianes. However as I indicate in Chapter 2, issues to do with indigenous Tagbanua communities of the Calamianes are heavily politicised. This politicisation, the long history of suspicion between many Tagbanua communities and outsiders, and the presence of newer guidelines requiring researchers to work closely with the National Council on Indigenous People, meant that the practical aspects of doing fieldwork with these communities were always going to be difficult to organise, and any research on them would have faced significant delays. I therefore avoided any explicit study of the Tagbanua communities, although many of the households I visited outside of Esperanza did identify themselves as Tagbanua.

The research methods that I adopted will be familiar to any practitioner of social anthropology, and were based primarily on informal interviews and observations. Early on during my fieldwork, I conducted a household survey of 70 households. Respondents were the husband and/or wife of the household. The questions focused on basic demographic data and also related to fishing and other livelihood practices. Apart from the data that I gathered, much of which was complemented by census data, the survey was a good opportunity for me to introduce myself to many of the people in the community, explain what I was doing there and develop a measure of rapport. As with all of my research in Esperanza, these interviews were conducted in Tagalog. Some more formal interviews were recorded; mostly, however, detailed notes were taken.

Conducting detailed life histories was one of my primary sources of data. Much of my most valuable data from Esperanza derived from informal conversations in the afternoons, as residents stood around socialising, playing basketball or volleyball, gossiping, or sorting through the fish catch. Similarly, when many fishermen were not out fishing they were busy at the local karaoke (*videoke*) house, which was another site of much informal conversation. Other ways in which I learned valuable information came through basic methods of observation, by accompanying fishing household members to the market and observing fish sales on the beach, going on fishing trips with different types of fishers, observing public meetings, participating in public and private social events, and going to church services. People from some households became more familiar than others, of course, and a great deal of information came in particular from those in the house where I was living. This household contained one of the more successful commercial fishing families in Esperanza.

Much of the material in this book deals with politically sensitive issues that are highly contested; themes of illegality and corruption pervade the text. I have used pseudonyms when referring to individuals and specific place names throughout the book, except for public provincial political figures and larger urban spaces like Coron town. While in some parts of Palawan and the Philippines the use of destructive fishing methods such as cyanide and dynamite is quite open (Galvez et al. 1989; Russell and Alexander 2000), in the Calamianes it is definitely conducted far more secretively. Because of the extreme sensitivity of this issue, I did not attempt to document in any detail the practice of illegal fishing. Instead, more importantly with regard to the argument I make in this book, I looked at the way illegal fishing was discursively constructed and contested. I heard a great many rumours and allegations against various specific individuals in Coron, but it is not the point of this book to provide any sense of an investigative report about who is responsible for illegal fishing. The only people that I refer to as being involved in cyanide fishing, in Chapter 6, are a group of young men who lived in various residences around Coron. I should emphasise that at no point in this book do I refer to anyone in Esperanza as being engaged in cyanide fishing.

Similarly, I would make the point that I am not alleging any specific instances of blatant corruption in the book. Instead, I show how perceptions of corruption among local residents inform their responses to regulatory regimes. As Walley found with regard to similar issues of illegality and corruption surrounding the establishment of a marine park in Tanzania, 'there is no way to make sense of [fishers'] actions without addressing this issue' (Walley 2004: 26) of perceptions about corruption.

As I demonstrate in Chapter 4, local residents around Coron tend to blame most illegal fishing on outsiders, many of whom are transient fishers. While I was able to meet and talk with these fishers, I found that overall I was not able to get much satisfactory data because of their high level of mobility. Understanding more about transient fishers would contribute greatly to overall understandings of fisheries in the Philippines.

Another key aspect of my research was interviewing relevant informants based in Coron town. One particular informant was well connected and extremely helpful in introducing me to members of the municipal council and the Calamianes Live-Fish Operators Association (CLOA), as well as other fresh fish traders. Through these introductions early in my time in Esperanza, I was able to conduct regular and detailed interviews in formal and informal settings with these 'powerbrokers' of the marine resource policy process, and attended numerous meetings.

By way of personal introduction, I formed close relations with certain members of the tourism industry, which mostly involved sitting around talking to the tourism operators at the local tourism establishments in town and on occasion going for a dive with them. I was able to follow their discussions on local issues, and observe their interactions with local actors. In a similar manner, I was able to meet representatives of all the conservation organisations working in Coron and to observe some of their project activities. Attending various project meetings took me to different locations throughout the Calamianes. On several occasions, I also travelled to Puerto Princesa City, the provincial capital of Palawan, interviewing government officials, NGO workers, live fish traders and fishers. In Manila, I was also able to interview some of the live and fresh grouper exporters at their aquariums, and to observe the process of sorting the fish when they arrived by boat from Palawan.

Ethnography and Cross-Disciplinary Dialogue

As I intend this book to be of interest to people from diverse backgrounds, here I briefly address some of the issues related to writing ethnography, and the general methodological approach that I have adopted here. I strongly feel that if we anthropologists want to do more than just talk to each other—if we want to contribute to debates about public policy—then we need to make more of an effort to explain our research and our methods in a clear and credible way.

I have not adopted the positions taken by the 'Writing Culture' school (Clifford and Marcus 1986) and some of the more extreme post-modern writers, such as the dialogism of Dwyer (1987) or the 'evocation' of Tyler (1986), which were developed to address concerns over ethnographic authority. The 'multi-sited ethnography' advocated by Marcus (1995) is a somewhat more useful way of overcoming some of the limitations of single-sited research.[3] I have explained in the preceding discussion how I had various field sites, and that my argument is based on experiences grounded primarily in Esperanza and Coron town, but also included other locations in the Calamianes, Palawan and the Philippines. And while for anthropologists this has become more the norm over the past several decades and is perhaps simply stating the obvious, it ought to be noted that this is not simply an ethnography of the 'exotic' customs and practices of an isolated village. While my book is in many ways a typical village-based study, I instead analyse how local 'traditional' ideas and 'modern' practices of environmental management are deeply intertwined. In saying this, my point is not to criticise village-based studies, or those that focus on 'traditional' anthropological topics, but rather to emphasise that these are not the only sorts of topics anthropologists can, and do study—a point not always appreciated by policymakers and other scientists.

In 1973, Geertz wrote that 'I have never been impressed by the argument that, as complete objectivity is impossible in these matters (as, of course, it is), one might as well let one's sentiments run loose' (ibid.: 30). Certainly, it is important not to let the particularities of the fieldwork experience overwhelm any sense of scientific objectivity based on the careful and thorough analysis of evidence. However, as Li argues, '[c]areful study of specific conjunctures—the kind of work conducted by anthropologists and social historians, among others—opens a space for theoretical work of a kind that is rather different from that of scholars

---

3 Although Marcus intended the term 'multi-sited' to refer to global cultural formations, not only to 'two villages', global formations such as 'conservation discourse' were not the prime focus of my research.

engaged in the immanent critique of theoretical texts, or the production of general models' (Li 2007: 30). Ethnography, from this perspective, is undeniably messy, inductive, imperfect and hermeneutic, yet still able to produce valuable analysis and theory.

More broadly, my methodological perspective draws on the work of Flyvbjerg (2001). He has argued that because universal laws cannot be discovered in the study of human affairs, concrete context-dependent knowledge is more valuable, and so the value of the case study should not be underestimated. Case study research offers the ability to explore issues of values and power in a way that more quantitative forms of analysis are unable to do.

I would argue that the reality that ethnographic knowledge is partial and ultimately subjective should not preclude its authority as a form that can produce uniquely situated perspectives. I have tried in this book to produce such a text; one that acknowledges its limitations, but one that is also based on lengthy and diverse interactions with a range of informants that has provided a grounded understanding of local life and culture.

Outline of the Book

The book can be roughly divided into two parts. Chapters 2–4 set out the context and content of the discourse of the poor moral fisher. Chapters 5–7 then shift the focus from perceptions to practices: these chapters explore how the discourse is expressed in various policy and everyday contexts, and how it has come to influence outcomes relating to marine resource use in the Calamianes.

Chapter 2 provides a detailed background to the rest of the book by examining the historical context through which Palawan has developed as a resource frontier. Conservation, commercial resource extraction and tourism are three forms of resource use contested at multiple scales in Palawan. The chapter analyses each pattern of resource use and the negotiations at three geographical scales: the provincial level; the Calamianes Islands; and within the sitio of Esperanza.

Chapter 3 provides further context by offering a detailed snapshot of the economic, class and status relations in Esperanza, focusing in particular on the social relationships involved in the various fisheries. Here, I describe in detail the economic patterns of the four primary fisheries of Esperanza, emphasising the role of personalised economic relationships.

Chapter 4 narrows the focus to describe the local understandings of the relationship between poverty and the environment that is the main theme of the book; an understanding represented in the discourse of the poor moral fisher. I show that this discourse depicts fishing as environmentally harmless and that it is closely tied to poverty. Through a detailed ethnographic description focused on the perceptions of fishers about fishing and their everyday lives, I show how this discourse relates to local ideas about fairness and pity.

Chapter 5 reveals how fishers used the discourse of the poor moral fisher during the implementation of a series of marine protected areas (MPAs). Fishers argued that implementing MPAs in their fishing grounds would be unjust, unless they were compensated from the profits of tourism. Here, fishers were able to manipulate these MPAs during planning and implementation, aiming to capture social and economic benefits, while ensuring to minimise their impacts on resource use.

Chapter 6 builds on the previous chapter by focusing on how certain fishers expressed their resistance to MPAs by continuing to fish within them. I found mostly young men participating in this form of fishing and analysed their resolve in terms of their particular economic and personal values. These values inform the practice of high-risk fishing, the need for rapid social and economic empowerment, and a desire to ultimately move out of the fishing sector and out of poverty.

Chapter 7 details the way in which fishers were able to participate in the process that brought about the rejection of regulations to reform the live fish trade by using the discourse of the poor moral fisher. Calamianes fishers argued that imposing a closed season, in particular, would greatly increase poverty, and ignored the 'real' problem of dealing with illegal fishing. In this chapter I analyse how this rejection of the regulations, and the discourse of the poor moral fisher, are deeply embedded within beliefs and attitudes about political life in the Philippines. The concluding chapter then draws all these themes and ideas together, exploring in greater depth the discourse of the poor moral fisher, and the implications that it may hold for alternative policy models for marine resource management in the Philippines.

2. Resource Frontiers: Palawan, the Calamianes Islands and Esperanza

Welcome to Rizal—the last frontier of Palawan, the country's last ecological frontier…. Mt. Mantalingahan's unspoilt beauty made more exhilarating by crystal clear rivers and miles and miles of virgin forest serve as home to varied, rare and endangered species of flora and fauna.

To those who have experienced its largely unspoilt natural and cultural treasures, Rizal is truly worth the visit, again and again. Its people, representing a diverse combination of different ethno-linguistic groups, have been a showcase of peace and unity. Working in harmony, they have been able to harness the untapped resources in them, and directing these into worthwhile courses of action geared towards the common goal of development.

Rizal is well on its way to becoming a major area of sustainable growth and development. With the support of its people, the Municipality shall move onward to more efficient utilization and conservation of resources through more responsive governance (mayor of Rizal, southern Palawan, 2006).

This quote by the mayor of Rizal from a tourist pamphlet exemplifies some of the tensions and contradictions underlying understandings of the frontier in Palawan. Here, a fundamental tension between development and conservation is apparent. I have considered it worth quoting at length because it provides an excellent snapshot of how the nebulous concepts of development and conservation are constantly confused and strained in Palawan.

While these comments could easily be simply dismissed as an example of the sort of political doublespeak for which politicians are so often criticised,[1] I prefer to view them as a valuable indication of some of the existent tensions over how to engage with the natural resources of the province. In the mayor's comments, conservation and development are seen as good things that are necessary and a common goal, yet the tensions and contradictions between them are not articulated. While the source of Rizal's uniqueness and value is its 'unspoilt natural and cultural treasures', the overall theme is the need for progress and development. These sorts of ambiguous messages are also in other promotional pamphlets, road signs, billboards and political speeches throughout Palawan.

---

1   And especially those in Palawan with reference to the concept of 'sustainable development'.

Then Governor Reyes, (2000–10) for example, would cite the need for 'Palawan's development' and speak of 'linking our growth areas to centres of trade', then in the next line talk of the need for 'environmental conservation and protection'.

Contested notions of the frontier form the setting for the specific arguments about poverty and the environment that I develop in this book. Palawan is a province marked by its abundance of natural resources, and has long been considered as a social and resource frontier. This notion of a frontier however, encompasses various interests that are frequently in tension with each other. Commercial extractive processes, conservation and related tourism activities are the most visible modes of resource use that enact contradictory notions of how the frontier should be managed.

In this chapter I aim to provide a background to the book by exploring some of the ways in which these conflicting patterns of resource use have developed historically. The chapter shows how different interest groups contest the ways in which natural resources are managed and exploited at the provincial, regional/municipal and *barangay/sitio* levels. I focus in the first section on how the province of Palawan has become a frontier. Secondly, I scale down to the region my book is focused on, the Calamianes Islands and the municipality of Coron in particular. I examine how this area is a marine resource frontier. Thirdly, I give some background information and detail how these issues have played out in Esperanza, the community in which I did my fieldwork research.

Palawan

Defining the Frontier

The term frontier has been used to describe various locations and cultures. Perhaps most famously, in the United States, Frederick Jackson Turner classified the regions just beyond the belt of non-European settlement as frontier lands. Lying between the 'wilderness' and the urbanised or settled areas, the frontier was 'the outer edge of the wave—the meeting point between savagery and civilisation' (Turner 1996 [1920]: 3). For Turner, the frontier was the zone that contributed to the making of the nation. He argued that the origins of American vitality, democracy and culture generally lay in the life of the frontier.

Since then, other academic and popular writers have adopted and used the term in various ways. In particular, many accounts have written about frontiers of capitalism (Tsing 2005). From this perspective, the frontier has been assumed as the region in which capitalism has yet to penetrate. 'Resource frontiers' were originally described by Friedmann (1966) as 'peripheral' zones of new settlement, such as has occurred in regions of the Amazon basin. Other writers have used this term to describe the ways in which regions have been progressively settled, and their natural resources extracted (Hyndman 1994; Brookfield et al. 1995). Brookfield et al. (1995), for example, show how Borneo and the eastern peninsular regions of Malaysia have been incorporated into the two countries of Indonesia and Malaysia as frontiers for particular natural resources; supplying oil, timber and other products.

Using the word 'frontier' to describe Palawan generally or without reference to specific resources can be problematic. For one thing, as McDermott (2000) pointed out, the term implies that 'expanding states, markets and migration have only recently reached its shores. In fact, translocal factors of incorporation have long shaped the course of social and environmental change on the island' (ibid.: 77). In the Calamianes Islands, for example, Indigenous Tagbanua were trading swallow's nests with the Chinese hundreds of years ago (Wright 1978: 56). Such a broad conception ignores the fact that Palawan is not a frontier for the various indigenous groups that have lived there for many years—the Molbog and the Pal'awan in the south of Palawan, the Batak in the north, and the Tagbanua in the central mainland and the Calamianes Islands (Eder and Fernandez 1996).

A different conception of the frontier looks at how the very idea of the frontier is constructed and enacted by various competing groups. Tsing (2005: 29) has pointed out that '[m]ost descriptions of resource frontiers take for granted the existence of resources; they label and count the resources and tell us who owns what'. Instead of using this approach, Tsing argued that resource frontiers, such as the Meratus Mountains in South Kalimantan where she has worked, are 'scale-making projects' (ibid.: 57–8); they enact particular visions of the landscape. Frontiers, from this post-modern perspective, are 'imaginative projects' (ibid.: 68) that construct the environment as wild and untamed, and the ways that people imagine frontiers are just as important as any objective notions about the frontier.

I do not intend to provide a rigid definition of the term frontier here, because by using Tsing's perspective on the frontier, I concentrate instead on how the frontier is perceived and enacted by various groups. Although Tsing did not define it as such, this approach closely resembles that of post-structural political ecology. In this sense, I refer to Palawan as a frontier because of the ways that different interest groups perceive it as such. I focus on the role of natural resources in the ways that the frontier is enacted by these competing interests. Here, I look at how the frontier is simultaneously an economic and a political space at the three geographical scales of the province of Palawan, the region of the Calamianes and the sitio of Esperanza.

Migration and Resource Extraction

Migration to Palawan has been a key driver of environmental and social change since the beginning of the twentieth century. In 1903, the total population of the province was 35 369 (Eder 1999: 24). By 2007, it was 892 660 (NSO 2010). During the early part of the century, migration was predominantly from Cuyo, a small island that is part of Palawan Province but lies in the northeast Sulu Sea, between mainland Palawan and Panay (see Map 1-2). Eder and Fernandez (1996: 7–11) relate that during this period, Palawan was not as important as a settler destination as other Philippine frontiers such as Mindanao. After World War Two however, Palawan experienced a large surge in migration that has continued. Migrants since this period, and especially since the 1980s, have predominantly come from the islands in the Central Philippines known as the Visayas (ibid.: 8). While there were some state-sponsored settlement schemes, such as that in Narra in Central Palawan, the majority of migrants to Palawan arrived spontaneously. This was mostly due to overpopulation, environmental degradation, and the prevalence of highly exploitative class relations and political unrest on Visayan Islands such as Cebu, Leyte, Samar, Negros and Bohol (Eder 2008: 37). I describe this process in more detail later when I discuss migration to Esperanza.[2]

The frontier more recently has been an ambiguous and contested notion in Palawan. As Eder and Fernandez have observed:

> … as the island and its peoples have become increasingly integrated into wider economic systems, the competition to control and exploit (or, more recently, to preserve) Palawan's resources has increased apace. Such competition today is very fierce indeed, and what may be called 'the politics of natural resource use' dominates both local, provincial, and national government development planning and the everyday lives of the island's residents (Eder and Fernandez 1996: 6).

The tension between development and conservation in Palawan can be seen more specifically in practices of commercial resource extraction, conservation and tourism.

Commercial resource extraction has long been an important economic activity in the province. Logging was dominant for many years in the post-war period, and during the 1980s it intensified with the activities of Pagdanan Timber Products. Owned by a businessman who had originally made his fortune exporting logs from Indonesia, Pagdanan Timber Products annual revenue by 1993 was

---

2   Seki (2004) has analysed in great historical detail such migration from Cebu Island in the Visayas to community settlements in Palawan.

US$24 million—equivalent to three quarters of the total income of the people of Palawan, and 24 times the provincial government's annual budget (Broad and Cavanagh 1993: 45).

The environmental movement waged a high profile campaign against Pagdanan during the 1980s (Broad and Cavanagh 1993: 39–55; Eder 2008: 31–2). Threats and intimidation against environmentalists were common in this period, but the movement was eventually successful. Commercial logging was banned in 1992 with the passing of *The Strategic Environmental Plan (SEP) for Palawan Act* (Republican Act (RA) 7611). At the time of fieldwork, mining, oil, gas and fisheries were important resource extraction activities. With a change in the mining laws to allow fully foreign-owned companies to operate in the Philippines from 2004, the Arroyo administration strongly encouraged mining. Palawan now has several operating mines and many more in stages of exploration, mostly in the south of the province. Gas and oil, mostly offshore to the northwest of the mainland, also provide considerable revenue for the province.

Much of the recent popular identity of Palawan as a pristine wilderness has been shaped by the enactment of the SEP (see Esguerra 1999). Recognising what it declared to be the unique environmental heritage of Palawan, the law aimed at managing the natural resources of the province in a sustainable manner. The main strategy was to divide the entire province into different zones that would regulate what activities could or could not take place within them. Core zones, for example, banned all human extraction of natural resources, while tribal ancestral zones were aimed at giving indigenous groups of Palawan increased levels of control over the land they occupied. The Palawan Council for Sustainable Development (PCSD) was created to implement the SEP. Significantly, a complete ban on commercial logging was put into place at the outset as part of the law.

NGOs, Conservation and Social Justice

Since the 1980s, a significant environmental movement has mobilised in Palawan. A host of national NGOs such as the Environmental Legal Assistance Center (ELAC), Tambuyog and Haribon has worked on projects throughout the province. Larger international NGOs such as Conservation International (CI) and WWF, the Conservation Organisation (WWF-Philippines) have also initiated and been involved in the management of several large-scale conservation projects. Partly because of these activities, Palawan now has two of the eight UNESCO World Heritage Sites in the Philippines. The origin, nature and activities of these environmental NGOs have been explored in detail by a host of academics including: Austin (2003); Bryant (2005); Austin and Eder (2007); Dressler (2009); and Novellino and Dressler (2010).

One of the most significant features of these NGOs, according to Austin, is their link to ideals of social justice. She points out that the leaders of many environmental organisations in Palawan were involved in human rights NGOs established during the time of Marcos. In contrast to other more preservationist forms of conservation, Austin (2003: 328) argues that the community-based coastal resource management projects she studied emphasise social justice as much, if not more than resource conservation. For many of these NGOs, the goal is to ensure stronger and more equitable rights of access to natural resources for local communities. For Austin, community-based coastal resource management 'is not just a coastal management program, nor is it a conservation/protectionist program; it is a social movement to allow fisherfolk the rights to livelihoods and food that was once plentiful in their communities' (ibid.: 53). Similarly, Bryant (2005: 69) argues that 'the pursuit of social justice and environmental conservation has tended to converge over time in the Philippine NGO sector ... it is the rare environmental NGO that does not ... recognize social and political preconditions for sustainability'. While several major international conservation NGOs such as CI and WWF have strong presences in Palawan, a related distinctive feature of the conservation movement in Palawan has been the development of what Austin (2003) terms 'meso-level' NGOs, staffed and headed by Filipinos.

A great deal of academic literature on Palawan has examined the activities of this conservation movement. Austin and Eder (2007: 365) assert that 'the environmental movement on Palawan, along with stronger roles for NGOs, has resulted in both better environmental protection and improved well-being for local peoples'. They highlight the 'hybrid' nature of Palawan's NGOs, arguing that many such NGOs successfully combine their linked goals of social justice and environmental advocacy. Lawrence (2002) also provides an optimistic account of a conservation initiative in Malampaya Sound, observing that it has fostered significant opportunities for local social change.

Such interpretations of the environmental movement in Palawan are contested. For Novellino (2007: 82), '[t]he old, strictly punitive protectionism is now being replaced by equally dangerous "community-based" forest management programs'. He declares that Western conservation in Palawan 'has the effect of ideologically disempowering indigenous communities, while jeopardising their livelihood patterns' (Novellino 2003: 172). He further argues that different moral beliefs among NGO workers/conservationists and their indigenous counterparts, as well as persistent state bureaucratic requirements, facilitate neither cultural nor environmental sustainability. Bryant contends that some of the NGO-led conservation initiatives in Palawan tend to reproduce processes of governmentality in their interactions with local communities. He views such NGOs as 'agents often (but not always) keen to empower the poor but who frequently serve to extend political rationalities of control and surveillance to hitherto "marginal" people and biota' (Bryant 2002: 286). Dressler (2009), in his analysis of the divisive role of the Subterranean River National Park, argues that the impacts of the initiative have exacerbated the ethnic and class differences between paddy and swidden farmers (see also Novellino and Dressler 2010).

My purpose is not to argue whether the conservation initiatives are 'good' or not, but these activities, combined with modes of resource extraction, provide the backdrop to perceptions of the poverty-environment relationship that form the focus of my analysis. Here I aim to emphasise merely the ways that conservation activities in the province have become institutionalised, and the heterogeneity of such activities.

Tourism

Tourism in the Philippines has been heavily promoted in recent years. *The Tourism Act of 2009* (RA 9593) formalised tourism promotion in law. In Palawan, tourism has been closely connected to the perception of the province as an ecological paradise, with government promotion focused on ecotourism and the natural features of the province. Indeed, for many years the official tourism website of Palawan featured a message from the then governor, who declared that '[c]onsidered as the Philippines' last ecological frontier, our pristine and unique natural resources and attractions continue to awe visiting tourists both domestic and foreign' (Palawan Department of Tourism 2007). Like this chapter's opening quotation from the mayor of Rizal, this message embodies contradictory images stating that: 'Palawan is a place where diverse activities may be held and different program [sic] that leads us to the continuous success of our conservation efforts as we move forward in the travel trade and surge toward economic development' (ibid.). Tourist blogs and websites related to Palawan attest to the region's unique setting, connected to notions of peace, tranquility and natural beauty.

Apart from a downturn in 2001, following the events of 9/11 in the US and a kidnapping incident involving an extremist Muslim group at the Dos Palmas Resort, tourism in Palawan has been increasing steadily. Provincial statistics up to 2004 state that arrivals reached a peak of 204 834 visitors in 2004 (Palawan Department of Tourism 2006). The most commonly visited destinations were areas of natural beauty: an underground river complex, island hopping expeditions in Honda Bay, and the beaches of El Nido. I will discuss shortly how Coron and the Calamianes Islands in particular are conceived of as a frontier for tourism, that is, as an area ripe for tourism investment and development.

Conservation initiatives, related tourism developments and resource extraction are three versions of the resource frontier that apply to Palawan. All three versions conceive of Palawan as a frontier, a place where natural resources are available for exploitation, but in different and often contradictory ways. I turn now to the Calamianes where these issues also play out, specifically with regard to marine resources.

The Calamianes Islands

Background to the Calamianes

> Coron. The next big thing (Slogan on t-shirts worn by Coron municipal officials at the Baragatan Provincial Festival, 2006).

> Tall rocky mountains. White sandy beaches. Exotic islands. Rich seawaters. A whole treasure trove of minerals and natural wonders. These are the things that crop to mind whenever one thinks of Coron (Palawan's then (and former 1986–87) Vice-Governor, Dave Ponce De Leon, Municipality of Coron, 2002).

The description of Coron—the largest town in the Calamianes Islands—as 'the next big thing' reflects a widespread feeling among people in the region that it is in the midst of great change and development. The second quote by Vice-Governor Ponce De Leon indicates the equally widespread perception that Coron's distinctive features are its natural resources. Like the quote from the mayor of Rizal at the beginning of the chapter, the comments by the vice-governor bring to light some of the tensions and contradictions associated with understandings of natural resources in Palawan. The phrases 'white sandy beaches', 'exotic islands', and 'tall rocky mountains' emphasise beauty and perhaps the possibilities of tourism. 'Rich seawaters', 'a whole treasure trove of minerals', on the other hand, indicate the potential for commercial benefit. Such incongruous perceptions of the resource frontier can be seen in the ways that different modes of economic activity are contested in the Calamianes. After a short general background to the region, I focus on the historical development of fisheries, dive tourism and conservation.

The Calamianes are a group of several hundred islands lying off the north of mainland Palawan, divided into four municipalities with their respective population sizes in 2007: Coron 40 007; Culion 17 194; Busuanga 19 066; and Linapacan 11 688 (NSO 2010). Coron, the municipality (as indicated on Map 1-2) where I was based, has a land area of 94 952 hectares equivalent to around 950 square kilometres. Somewhat confusingly, the municipality of Coron includes the island of Coron, but Coron town and most of the rural barangays are located on the eastern portion of Busuanga Island (see Map1-1).

The municipality of Busuanga administers the rest of Busuanga Island. In 2007, there were 8577 people in the five barangays of the major town, Coron; while a further 31 500 lived in the 18 rural barangays.

The indigenous people of the Calamianes are known as the Tagbanua. They are linguistically distinct from the Tagbanua of the interior of mainland Palawan (Fox 1982). They are recognised in the Philippines as an indigenous cultural minority, and in 1998 they were given control of the land and waters surrounding Coron Island, under *The Indigenous People's Rights Act of 1997* (PAFID 2000; Bryant 2002). They were struggling to gain Ancestral Domain recognition of further parts of Coron municipality when I was living there and the issue of land rights for the Tagbanua was a heavily politicised issue; similar to indigenous issues voiced elsewhere in the Philippines. In particular, questions about indigenous authenticity are commonly raised in the Philippines (Scott 1982: 28–41; Hirtz 2003), and the Calamianes were no different when I worked there. Tension also circulated over the growing numbers of tourists travelling to Coron Island. This island boasts prominent tourist attractions such as the Kayangan and Barracuda Lakes. Some in the tourism industry resented what they saw as the Tagbanuas' efforts to gain control of the economic benefits of tourism without maintaining or investing in such attractions. Numbers of other foreign and Filipino residents around Coron municipality looked on the Tagbanua disparagingly.

The pattern of migration to Coron and the Calamianes mirrors the pattern that occurred throughout Palawan more broadly during the twentieth century (Eder and Fernandez 1996). During the first part of the century, migration occurred mostly from Cuyo. Migration from the Visayas intensified during the 1970s and 1980s, especially fishermen 'who first appeared in Calamianes as crewmen on fishing vessels out of Manila' (Wright 1978: 66).

Coron municipality now has a high level of linguistic and ethnic diversity relative to the rest of the country. Tagalog is the lingua franca, but Tagbanua and Cuyonon are also widely spoken among Tagbanua and Cuyonon residents. Tagbanua people live on settlements on Coron Island as well as other rural areas of the municipality. Visayan and Cuyonon settlers also spread throughout the rural areas but many also live in town. Businesses in town are operated by a range of groups, including Cuyonon, Visayans, and other migrants from Manila or elsewhere in Luzon, Chinese-Filipinos, and various foreigners from a range of Western countries. Politically, four or five extremely influential families have maintained strong influence since Coron was registered as a municipality in 1902. These families are involved in a 'dynastic' form of politics, however, their influence falls short of a monopoly on power, and other individuals and families have participated in the political arena throughout the history of Coron.

Coron was physically connected to the rest of the Philippines during 2006–07 by an airport that serviced daily passenger flights in small planes and cargo planes carrying live fish (see Plate 2-1). This airport was upgraded in 2008 with capacity to handle 50-seater planes. In 2006–07 Coron was also serviced by two large passenger ferries from Manila that stopped on their way to Puerto Princesa, and several smaller mixed cargo and passenger vessels. Although the roads in urban Coron are sealed, roads that circle Busuanga Island and those connecting inland settlements were in 2006–07 for the most part unsealed. As elsewhere in the Philippines, the Roman Catholic Church has the largest following, but other denominations are represented by the Filipino Catholic Church of Christ sect, Iglesia ni Cristo, and various Protestant Churches. There are also increasing numbers of Muslims who have migrated from southern Palawan and Mindanao.

Plate 2-1: Live fish caught in the Calamianes loaded for export at Coron Airport.

Fisheries

Fisheries have a well-established history in the Calamianes, closely connected with the continuing influx of migrants (Fabinyi forthcoming). The fishing grounds of the Calamianes have been long considered as a resource frontier for migrants. While the region has experienced various booms and busts since the end of World War Two, the trend more recently has been towards the operation of export fisheries. The scale of fisheries activities in the region has intensified especially since the 1990s with the development of the live reef fish for food trade.

In a review of Philippine fisheries in 1948, Herre noted that the fishing ground near the Calamianes, the South China Sea, was:

> a virtually untouched source which has been shown to be suitable for trawling over large areas. Vast, uncharted shoals, banks and reefs west of Palawan are reported to abound in tuna, bonito, scads, and a great variety of reef fishes. These rich fishing grounds await the development of a modern fishing fleet (Herre 1948: 278).

Japanese fishing vessels had been fishing Calamianes waters for some years, but the real boom in fisheries for the people of Coron began during the 1960s with the influx of many bag net or lift net (*basnig*) boats. These boats used a large bag net and fished at night with the aid of gaslights, attracting large numbers of anchovies (*dilis*). Many new arrivals to the Calamianes were able to join as crew members on these boats and owners rented their boats to local captains. Mostly, the boats would fish around Coron Bay with trips lasting for only three or four days at a time. During my fieldwork, residents nostalgically described this period as a time of plenty. It reached its peak in the 1970s: by 1976 there were 115 basnig boats owned and based in Coron (Baum and Maynard 1976: 23), but catches declined from the early-1980s, and in 2006–07 there were fewer than ten boats operating in the whole region.

During the 1970s, the Calamianes were also host to a wide range of commercial vessels based in Manila. At this time, Wright (1978: 48) stated that 'Calamianes is a major fishing ground and a transit point for the largest fishing companies which operate out of Manila Bay'. In particular, the Calamianes and Palawan more broadly were the primary sites of the now notorious *muro-ami* fishing vessels. These boats used a technique involving hundreds of swimmers, often children, who attached nets to the corals. They then pounded the corals with rocks attached to ropes, scaring the fish into the waiting nets. The biggest *muro-ami* fishing family, from southern Cebu, had its base of operations on Panlaitan Island in the Calamianes and fished the waters of northwestern Palawan and the South China Sea (Butcher 2004: 192–3; Fabinyi forthcoming).

From the 1970s until the time my fieldwork, various other commercial fisheries have developed; some have had good yields for a long time but others have disappeared as stocks quickly depleted. Various trawling operations, especially the Danish seine (the *hulbot-hulbot*), have been operated by local residents since the original basnig boom (see Ingles 2000). Other commercial fishing boats based in Coron have included boats targeting fusiliers, or boats targeting multiple species with hand-lines. Baby purse seines based in Manila and Mindoro operated in the region on a seasonal basis for many years. Apart from commercial fisheries, residents of the region also target fish using a variety of small-scale and subsistence techniques. A baseline survey of the capture fisheries in Coron Bay in 2004 (Cruz 2005) found that residents from the region used more than 30 gear variations, under seven major categories of which gillnets and hook-and-line were the most popular.

Since the early-1990s, one fishery has emerged as dominant in the Calamianes: the live reef fish for food trade (see Plates 2-1, 2-2, 2-3).[3] It has steadily increased, responding to growing international demand from the rapidly industrialising regions of Asia. By 2005, live fish export value in the Philippines had increased from US$7.2 million per year between 1991 and 1998 to a yearly average of around US$11 million between 2001–03 (Pomeroy et al. 2005: 16). The live fish trade in the Philippines began in Guiuan in southern Samar during the 1980s, and, following a pattern similar to other source countries in Asia, moved to new regions as stocks became depleted. By the 2000s, the live fish trade was concentrated in Palawan, which in 2002 was estimated to produce 55 per cent of the country's live fish (Padilla et al. 2003: 7). Within Palawan, Coron has historically been a focal point for the trade (Fabinyi and Dalabajan 2011).

Plate 2-2: Leopard coral grouper atop a float representing Coron at a provincial festival held annually in Puerto Princesa.

---

3 The term 'live fish trade' is often used to refer to both the marine ornamental trade (see Plate 2-4) and the live reef fish for food trade. While fish for the marine ornamental trade were collected in the Calamianes, the extent of this trade was far less than the live reef fish for food trade. Hereafter, when the term live fish trade is used it refers specifically to the live reef fish for food trade.

Plate 2-3: Leopard coral grouper held in an aquarium.

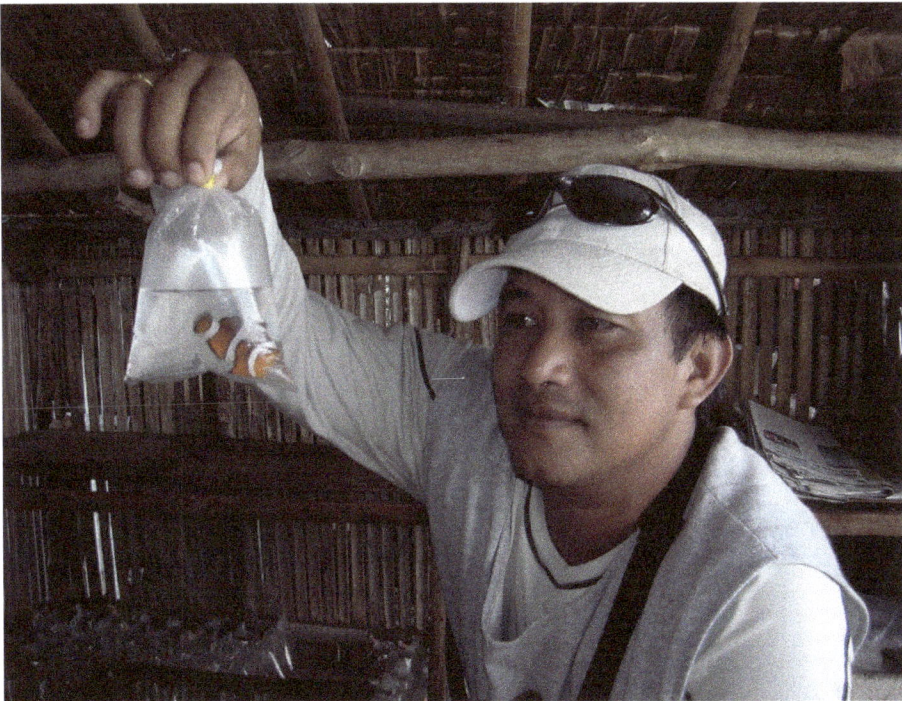

Plate 2-4: NGO worker holding a fish caught for the marine ornamental trade.

The live fish trade developed in the Calamianes during the early-1990s and by the end of the decade, 60–70 per cent of communities were involved in live fish collection (Padilla et al. 2003: 24). In the Calamianes, the live fish trade is virtually a one-species fishery, focused on *Plectropomus leopardus*, also commonly known as the leopard coral grouper.[4] These fish are kept alive during the fishing trip (which can last up to two weeks), before they are transferred to a holding facility in Coron. After a short rest period, they are flown via Manila to Hong Kong, where many are re-exported to China.

Tourism

Tourism is a second way in which Coron is conceived of as a marine resource frontier. Seen as beautifully 'unspoilt' and ripe for development, Coron is being heavily promoted by municipal, provincial and national governments as a new tourist destination. The development of new resorts, accompanied by escalating land value and the increasing numbers of foreign and Filipino tourists all testify the view of Coron as a new Philippine tourist destination.

In 1978, Wright had as one of the themes of his Ph.D. in Geography the development of a plan to promote tourism in the Calamianes. He argued that the Calamianes possessed geographic, climactic and natural assets that had positioned it perfectly to take advantage of the growing tourism market. As he put it: 'Calamianes is pollution free. There is no overcrowding or traffic. There is no latent security hazard. Floods, landslides, tidal waves and earthquakes are not known here. The main typhoon zone is to the north. Calamianes is close to Manila, but little-known and unspoiled' (ibid.: 180). He advocated strong investment in tourism based on these features. Modelling his idea of tourism on regional locales booming at that period such as Pattaya in Thailand, his ideas about tourism were very different to the current rhetoric of ecotourism. Among other ideas, he advocated the development of a military club, a 'sunset' club at the beach, 'special' clubs for sailors and flyers, country clubs at the golf courses, and a helicopter pad and restaurant on the top of Mt. Tundalara, the highest mountain on Busuanga Island (ibid.: 216).

Through the 1980s, growth in tourism did not increase in the Calamianes as it did in other regions of the Philippines such as Boracay and Subic Bay. Instead, tourism eventually became linked with SCUBA diving mostly; offering 14 Japanese shipwrecks, bombed by US planes in 1944, around Coron Bay and Busuanga Island. For SCUBA divers who enjoy wreck diving, it is marketed as one of the most appealing locations in all of Asia. Dive tourism in Coron began to develop in the early-1990s and within ten years, there were eight dive

---

4   Also known as the leopard coral trout or common coral trout.

companies based in Coron town. Municipal government statistics on tourist arrivals in Coron were unreliable before 2008, and to compensate for lack of data, I found that talking to tourism operators who had been operating since the 1990s was a better way of obtaining a general understanding of the level of tourism. After a peak in the late-1990s, there was a significant downturn in arrivals after 2001, when the 9/11 attacks in New York and the Dos Palmas Resort kidnapping incident seriously damaged Palawan's tourism industry. Many dive operators were forced to leave. Since this time, and especially since 2008 with the opening of the new airport, the number of dive operators in town has increased again, and more are based in locations outside of Coron town with their own resorts.

More recently, tourism in Coron has diversified. Lodging houses, other than those associated with dive operators, have increased and tourism businesses are trying to promote an array of activities for tourists. These are mostly still outdoor-based attractions, such as 'beach hopping' (visiting multiple beaches), kayaking, yachting, fishing, and rock climbing. Importantly, the local municipal government in Coron elected in 2004 (subsequently re-elected in 2007 and 2010), made tourism a priority. After long negotiations, redevelopment of the Coron Airport finally began in June 2007, with high anticipation of larger numbers of international and domestic tourists. Improvements to achieve this included the concreting and expansion of both the runway and the apron. The new airport opened in November 2008, with then President Arroyo 'express[ing] elation' about the increasing numbers of tourists (Gov.Ph News 2008). Tourist arrivals to Coron have increased rapidly since the opening of the airport: a total of 13 980 arrivals passed through the airport in 2008 and by May 2009, 18 555 visitors had already been logged for 2009 (Palawan Times 2009).

One of the most potentially important tourism developments in Coron has been the high level of investment from members of the Boracay Group, a consortium based in the heavily developed resort island of the same name in the Visayas. During 2006–07 the group was acquiring resorts based around the Calamianes, and proposals were underway for the construction of several new ones. These investors saw the potential and opportunities in the beautiful beaches and islands of the Calamianes to create a 'Little Boracay', and this interest was noted by the then president in her 2007 State of the Nation Address (Arroyo 2007). Another high-profile tourism resort being developed is by the Singapore-based Banyan group, and has been showcased by the provincial and national governments as having the capacity to turn Coron into a 'new Phuket' (Calleja 2009).

When I interviewed local municipal officials about tourism, it was clear that many of them viewed the industry as a healthy, relatively risk-free alternative to fishing and farming. The mayor, for example, said that his aim was to promote tourism as the central economic activity during the reduced fishing season from

October to May, during the dry season. Many of his public speeches during 2006 were marked by references to Coron as a booming place, as evidenced by the building of new hotels and guesthouses. In 2008, then Palawan Governor Joel Reyes allocated for Coron more than ₱235 million[5] out of the total ₱966 million fund deriving from the Malampaya gas project (Villanueva Jr 2008). This was the largest share of any municipality in Palawan, with Governor Reyes arguing that this was justified because of the high tourism potential of Coron. Other officials in other municipalities of the Calamianes were just as optimistic about tourism, talking about the potential of the stunning beaches in some of the more remote islands. They argued that it was necessary to 'develop' the beaches in order 'to prepare for when the tourists come'.

Foreign investment is also channeled into the development of retirement properties. Following the trend throughout the Philippines of attracting older retirees from around the world, this form of investment has recently become more common in Coron, and property prices throughout the municipality are increasing rapidly. Small islands and coastal properties on the Busuanga mainland now routinely sell for millions of pesos. In nearby San Vicente, on the Palawan mainland, Eder (2008: 55) found that good locations were extremely expensive, and that 'more than half of San Vicente's prime beachfront property, both along the coast and on the various offshore islands, is under the ownership of foreigners and a handful of wealthy provincial politicians'. During 2006–07 in Coron at least half a dozen foreigners moved in and bought property. In other regions of the Philippines, such as parts of Negros, these trends are even stronger (Knudsen 2008).

Marine Conservation and Management

While the Calamianes have been technically subject to the same Strategic Environmental Plan (SEP) laws as the rest of Palawan since their inception in 1992, large-scale conservation initiatives were only introduced relatively recently. Conservation constitutes a third important, related collection of people and interests struggling to prioritise their own access to marine resources in the Calamianes.

In 1998, a CI team conducted a 'Rapid Marine Biodiversity Assessment' of the Calamianes and subsequently concluded:

> The results of the Marine RAP survey firmly establish the Calamianes Islands as a primary target for marine conservation, especially in view of its extraordinary diversity of corals, molluscs, and other organisms, rapidly diminishing habitats and over-exploitation of marine resources.

---

5   The currency referred to throughout this manuscript is the Philippine Peso (₱).

There is an urgent need for immediate action to prevent further degradation of the environment and to conserve a still relatively rich biodiversity (Werner and Allen 2000: 8).

Other reports have since confirmed the value of the marine environment around the Calamianes (FISH 2005). Despite the activities of CI and a few smaller NGOs, large-scale attempts at conservation and management have begun only recently. In 2005, a large-scale USAID-funded marine conservation project started its activities, and in 2004, a Japanese Government-funded 'Sustainable Environment Management Project in Northern Palawan' (SEMP-NP) directed its efforts to introduce various marine protected areas in the islands.

The SEMP-NP's goal was 'to conserve the precious environment and natural resources in Northern Palawan and provide alternative income methods through environmentally sustainable tourism for host community members' (Green 2004: 3). In order to do this, SEMP-NP worked with the Palawan Council for Sustainable Development (PCSD), the Department of Environment and Natural Resources (DENR) and other bureaucratic arms of the Palawan administrative and political landscape to create a number of MPAs on the most popular dive sites around Coron. Chief among these dive sites were the sunken Japanese shipwrecks, previously under joint management of the dive operators based around Coron. Another MPA was created on a coral reef dive site closer to town. By the end of 2005, the process initiated by SEMP-NP was in operation and tourists were paying user fees for entry to dive sites.

The FISH project was a large USAID-funded project that ran from 2003–10; heralded as the first large-scale example of ecosystem-based fisheries management in the tropics (Christie et al. 2007). The project site in the Calamianes was one of four target management areas in the Philippines. The overall goal of the project was to achieve a 10 per cent growth in fish stocks by 2010, through the following activities:

- promotion of coastal resource management as a basic service of local governments;
- strengthening of the coastal law enforcement program;
- establishing a network of MPAs;
- introducing fishing effort restriction measures; and
- implementing institutional capacity building and constituency building.

While people running the FISH project were not originally involved in the set up of the MPAs introduced by SEMP-NP, they were later providing technical assistance to the communities affected by the protected area designation. They attempted to develop many other MPAs at different locations throughout the Calamianes. They also held workshops with government officials on coastal management issues, and were advising on a range of policy and legislative issues such as reform of the live fish trade, and the introduction of compulsory registration for all municipal fishers.

Other NGOs working on marine conservation in the Calamianes during 2006–07 included the Environmental Legal Assistance Centre (ELAC), a Filipino NGO promoting the creation of MPAs and legal advocacy work, and the Marine Aquarium Council, operating to reform the marine ornamental trade. There are a range of other government institutions that are typically involved with any attempt at marine resource management, most notably the Department of Environment and Natural Resources (DENR), the Bureau of Fisheries and Aquatic Resources (BFAR), the Department of Agriculture, and the Palawan Council for Sustainable Development Staff (PCSDS). With such a diversity of local actors the conservation movement in Coron defies easy categorisation in terms of being either 'preservationist' or foreign. All permanent project staff in the Calamianes during 2006–07 were Filipino and the motives and aspirations of those involved in conservation work were diverse.

Like Palawan, competing interests see the Calamianes as a frontier. Marine conservation, tourism and commercial extraction are three related assemblages of interests, which contest contradictory versions of the frontier. I now move to a closer examination of how such issues figure in the daily lives of the residents of my primary research site, the *sitio* of Esperanza. I begin by giving some historical and demographic background before detailing the role that marine resources have had in Esperanza.

Esperanza

Historical and Economic Contexts

While a Spanish-Filipino family claim to have held title to the land from the early part of the twentieth century, they were absent from the area until very recently. Migrants from Waray-speaking parts of Northern Samar began arriving in Esperanza in the late-1960s. The Calamianes and Palawan more generally had a reputation at this time of bountiful marine resources, and a sparse population. During this period, there were only two Indigenous Tagbanua families living in the area of Esperanza and the coastline was only lightly settled. Almost all of these early settlers described their motives for settling as wanting to gain better economic opportunities through fishing. Migration from Samar continued in the 1970s with families from other parts of the country arriving, mostly from islands in the Visayas. During the 1980s, the rate of migration increased, and according to the latest government census of 2007, the sitio had 103 households with a total population of 529 people. While it is technically a sitio within the larger political unit of a barangay, its rural nature, large population and length along the coast make it more cohesive and independent than many other sitios. The barangay of which it is part stretches for several kilometres along the coastline. This area has a long history associated with fisheries in the Calamianes with many of the basnig boats in the 1970s based there and has as patron Saint Raphael; often depicted standing on top of or holding a fish.

Esperanza's historical appeal as a migrant destination means that it is now home to people from all over the Philippines. Notwithstanding the high diversity of people however, the distinctiveness of such groups is changing over time, as indicated by the birthplace of current residents obtained by a survey of 70 households comprising 352 people (Table 2-1).

Table 2-1: Birthplaces of Esperanza residents.

Birthplace	Percentage of total
Coron	51.4
Samar	14.2
Bohol	9.9
Cebu	4.5
Negros	4.3
Romblon	3.4
Other	12.9

Source: Author's calculations based on household survey data, 2006.

The fact that over half of those residents surveyed were born in Coron is an indication of the high proportion of younger residents. While the vast majority of household heads were born in other provinces, they settled in Esperanza and had children who by 2006 outnumbered the original migrants.

One significant consequence of this is that Tagalog is now the lingua franca of the community. While various dialects, mostly from the Visayan region, continue to be spoken among adults from specific regions, Tagalog is the first language for most children growing up in Esperanza. I did not obtain survey or detailed data on what languages other households speak within their homes, but from my experience, Tagalog was used almost exclusively within the households I visited and stayed. Regional dialects tended to be used when residents communicated with relatives living in their original Visayan communities. In relation to Eder's observation that 'when, and how ethnicity "matters" in the lowland Philippines is still an important problem deserving greater scholarly attention' (Eder 2003: 223, see also Eder 2004). I see the way that Esperanza is classified by outsiders as a community of 'Visayans' as far more important than internal ethnic differentiation of Warays from Samar, Boholanos, Cebuanos and so on.

In the Philippines, Visayans are frequently seen as a nomadic, sea-going people who are naturally drawn to the ocean. In Palawan in particular, they are viewed as particularly responsible for the widespread environmental degradation that has occurred in recent decades. Around Coron, and in town especially, the widespread view is that Visayans are poor fishers who cannot afford to do anything else. A few quotes serve to illustrate these perceptions.

> Be careful, they are Visayans! They are poor! (An older, urban Cuyonon woman warning me of the dangers of living in a Visayan village at the start of my fieldwork).

> They are like the Chinese. They are everywhere! (A conservationist from Manila, joking while visiting Coron).

> They are only above the Tagbanuas (A Coron resident discussing social hierarchies of different ethnic groups).

> They have a nomadic and destructive lifestyle (A conservationist referring to the reputation of Visayans as nomadic and their association with cyanide use).

This last quote is an example of a more widespread discourse (in the Philippines and elsewhere) that links 'nomadism' to environmental degradation (see Lowe 2006). Similar to how this term has been used to describe another mobile maritime group, the Bajo, the implications of continual, purposeless movement are inaccurate (Lowe 2006; Stacey 2007). Instead, with many Visayan fishers, seasonal or even permanent migration is more common (Zayas 1994; Fabinyi 2010).

Many indigenous Tagbanua residents were highly critical of Visayan migration to the Calamianes, arguing that Visayan fishing practices have had a massive impact on local livelihoods (Dalabajan 2000: 170). Leaders of Tagbanua communities, and frequently the NGOs that support them, have typically framed indigenous practices of resource use in direct opposition to those of migrants. They argue that while Tagbanua people have lived in relative harmony with their environment for hundreds of years, the problems of overfishing and coral reef destruction only began when migrants arrived in the Calamianes.

In his 2003 article, Eder discussed livelihood options for coastal residents in San Vicente on mainland Palawan. He critiques the popular notion that Visayan people are an ethnic group drawn to fishing and the life of the sea because of their cultural and ethnic background. He shows that their association with the sea in Palawan is due more to distinctive historical experiences, rather than any 'cultural calling' (ibid.: 218). He points out that Cuyonon and Agutaynen settlers migrated to San Vicente well before the earliest Visayan settlers, and that by the time Visayan migration intensified in the 1980s the price of land was prohibitively expensive. Visayan settlers therefore became drawn to fishing through economic necessity. He also offers an argument as to why we should consider fishing and farming not as separate, discrete strategies, but as intimately linked practices. He demonstrates the ways in which fishing and farming practices relate to each other, and points out that rarely are coastal communities simply composed of fishers or farmers (see also Dressler and Fabinyi 2011).

Notwithstanding the interconnectedness and variety of livelihoods among and within many coastal households in Palawan, Esperanza represents somewhat of an exception with its predominantly marine livelihood orientation. The heavy reliance of residents on marine resources is due to several factors. Firstly, the land around Esperanza, and indeed all around Busuanga Island, is drier than much of the Philippines, limiting agricultural options compared to other areas of Palawan. One of Busuanga's most important crops is cashew, which is particularly hardy and well suited to drier tropical climates. Secondly, there is no irrigation or freshwater source around Esperanza that can be used for agriculture (wells provide the water for washing and drinking for most families). Thirdly, and perhaps the most important reason driving reliance on marine resources is the situation regarding their precarious hold on land title. For several years, the residents of Esperanza have been involved in a land dispute with the original Spanish-Filipino family who claimed title of the land and who had returned to Coron with the intention of developing the land for tourism purposes. The protracted legal negotiations and insecurity of their land tenure mean that Esperanza residents have been unable to farm to any significant degree any of the nearby land.

Despite this strong reliance on marine resources, there are still significant land-based incomes to be made in the sitio. During the dry season (*amihan*), when fishing is less profitable, some residents work as labourers or tricycle drivers in Coron town. One better-off family was able to acquire some land on the other side of Busuanga Island (a gift from a foreign son-in-law) for them to work and provide some employment for other residents. Others maintain what is locally referred to as a 'sideline' or a 'Plan B', a secondary income from raising hogs and chickens, working as domestic help in Esperanza or in town, or running a general store (*sari-sari* store). As Eder (2003) points out, often these 'sidelines' are maintained within households that usually rely on fishing as their primary income source. Typically it is women who work these 'sidelines' while their husbands go out fishing.

A summary of income sources for Esperanza residents from the government census data of 2007 is given in Table 2-2.

Table 2-2: Occupations of people in Esperanza.

Primary Occupation	Number of people (n = 179)
Fishing	112
Labourer	15
Marketing	10
Other	42

Note: occupation categories give little indication of the occupational multiplicity that exists. Source: NSO 2010.

The census data also indicates that Esperanza residents obtained a total income for 2006 of ₱5 521 000, equivalent to an average of ₱54 663 and a median of ₱46 000 income. Of this total income, ₱2 531 100, or 46 per cent, was gained through fishing, with an average income of ₱31 637 and a median income of ₱30 000 for fishers. These figures are only approximate and obtained using on-the-spot questionnaires of fishers, not by detailed long-term accounting or observation. However approximate, they do point to interesting aspects of income distribution. That 63 per cent of workers nominated fishing as their primary income, yet only 46 per cent of total income was produced by fishers points to the fact that while fishing is a common occupation, it is often the least financially rewarding. The large discrepancy between average and median incomes is most probably a reflection of the inclusion of several Manila-based workers on much larger incomes in the survey sample.

The role of dive tourism and marine conservation for the residents of Esperanza has become far more prominent in recent years with the development of the Esperanza Marine Park; one of the MPAs developed by the SEMP-NP. This is analysed in detail in Chapter 5 when I discuss the impacts of the MPA and attitudes towards it; namely, there had been few employment opportunities and the money from user fees had yet to be distributed while I was in the field. Three residents worked as tourist guides on dive boats, and many of the carpenters had worked on the dive boats owned by the dive operators. Indeed, Esperanza holds a reputation as a stronghold of master boat-builders (Plate 2-5). Mr Rosario was one such boat-builder, an older man who during his time in Esperanza had taught his techniques to various apprentices, both from Esperanza and from other barangays. He had developed a reciprocal relationship with the owner of the biggest dive shop in Coron town, and his responsibility was to build and maintain the boats and engines of all the boats. Apart from the impact of the MPA and those few residents who worked in tourism-related industries, however, Esperanza's everyday economic engagements with marine resources are still primarily based on fishing. I turn now to the various fishing activities in Esperanza.

Plate 2-5: A carpenter working on his boat.

Fishing

The development of the Esperanza fisheries (as shown on Map 2-1) was typical of the pattern in the Calamianes I outlined earlier and is further detailed in Table 2-3 with the commonest fish caught provided in Table 2-4. The first major fishery people were involved with was the basnig fishery. Being new to the area and mostly very poor migrants, these people had little access to start-up capital and simply joined on as crew members on one of the many boats already fishing the waters of Coron Bay. The barangay captain at the time was a large figure in the basnig fishery and he owned most of the boats on which local Esperanza residents worked. Residents described this period as a time of plenty; fish were abundant and easy to catch. Fishing trips normally lasted only 3–4 days and were based within Coron Bay. As in other areas of the Calamianes, this fishery has now declined. By 2006–07 there were just three basnig boats from the barangay operating.

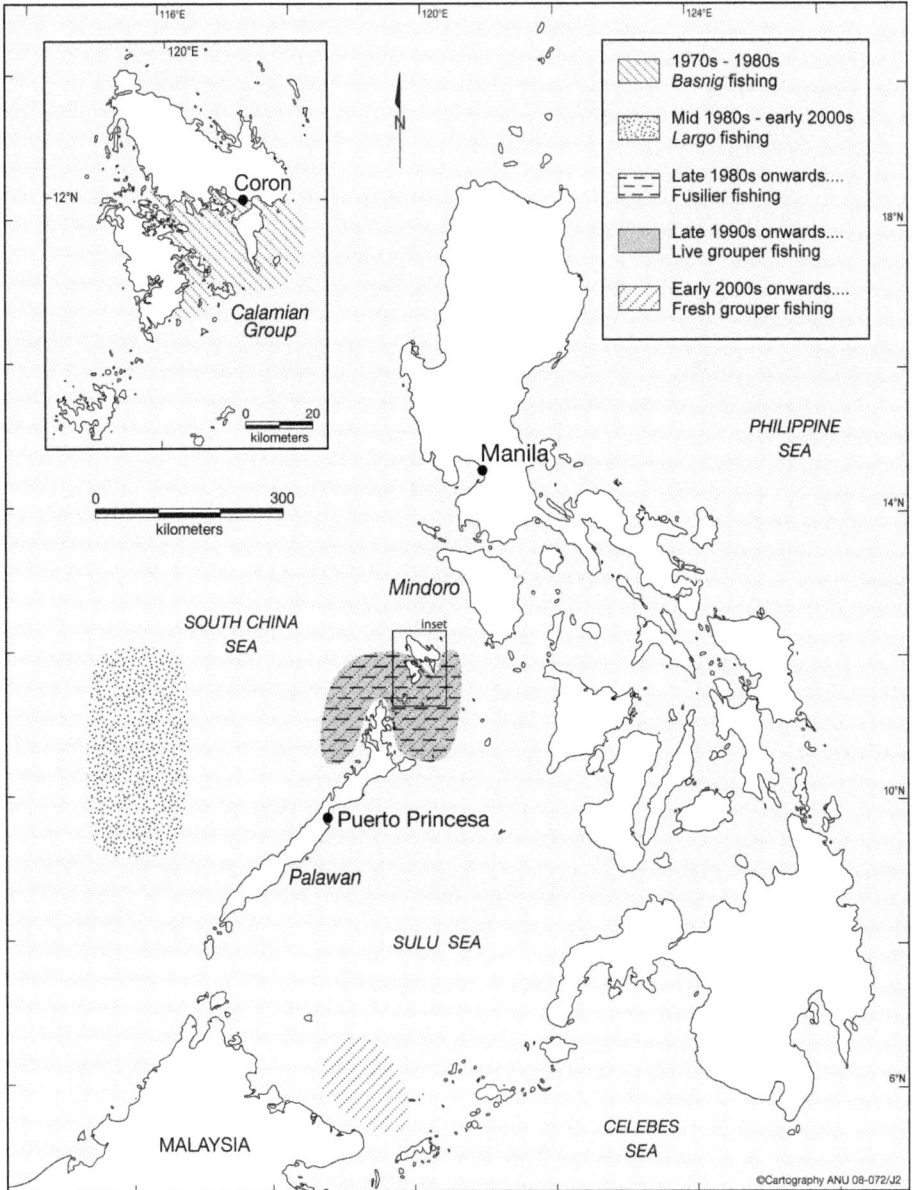

Map 2-1: Historical development of fisheries based in Esperanza.

Source: Cartography ANU.

Table 2-3: Historical patterns of fishing in Esperanza.

Fishery	Time period	Fish targeted	Gear	Location	Crew
Basnig	1960s–early 1980s	Anchovies, Scads	Lift net	Coron Bay	20
Largo	Early 1980s–late 1990s	Variety of large fish	Hook-and-line	South China Sea	20
Net fishing	1970s–present	Rabbitfish	Variations of gillnet	Coron Bay	2
Fusiliers	Late 1980s–present	Fusiliers	Hook-and-line with bait	Dibangan	20–30
Live grouper	Early 1990s–present	Groupers, Leopard coral grouper	Hook-and-line with bait and lure	Calamianes, Northern Palawan	4–6
Fresh grouper	Early 2000s–present	Groupers, Leopard coral grouper	Hook-and-line with lure	Dibangan, Sulu Sea	20–30

Source: Author's fieldwork data, 2006.

Table 2-4: Common fish caught in the Calamianes waters.

Local name	Scientific name	Common name
Alumahan	*Rastrelliger kanagurta*	Long-jawed mackerel
Bisugo	Family Nemipteridae	Threadfin breams
Dalagang bukid	*Caesio cuning*	Red-bellied fusilier
Danggit	Family Siganidae (smaller spp)	Rabbitfish
Dilis	Family Engraulidae	Anchovies
Galonggong	*Decapterus* sp.	Scads
Kanuping	Family Lethrinidae	Emperors
Lapu-lapu	Family Serranidae	Groupers
Mulmul	Family Scaridae	Parrotfish
Pusit	Order Oegopsida	Squid
Samaral	Family Siganidae (larger spp)	Rabbitfish
Suno	*Plectropomus leopardus*	Leopard coral grouper
Sulid	Family Caesionidae	Fusiliers
Tambakol	Family Scombridae	Tuna
Tangigi	*Scomberomorus* sp.	Spanish mackerel

Source: Author's fieldwork data, 2006.

As fishermen in Esperanza became more successful, they began to build their own boats. In response to the decline of the basnig fishery, during the early 1980s most fishermen switched to another type of commercial fishery called *largo viaje*. This method used heavy lines with multiple hooks, and targeted large, 'first-class' fish such as snapper, tuna, and trevally which were transported to markets in Batangas and Manila. Fishers were eventually forced to move to a fishing ground known as Reed Bank, far out in the South China Sea due to decline of fish stocks in Coron Bay. These trips lasted up to two weeks. Largo remained the dominant fishery in Esperanza until the late-1980s, when the increased price of fuel and other expenses forced fishers to look for other species closer to home. While largo continued to be used by most of the commercial operators during the 1990s, after this time it became a seasonal fishery and the fusilier fishery took over as the primary method used in Esperanza.

Fishermen modified their hook-and-line techniques in the early 1990s to target fusiliers in the area around Dibangan Island, a rich fishing ground southeast of the Calamianes. For most of the 1990s, these operators fished for fusiliers during the rainy season and for the first class fish in the South China Sea during the dry. By the turn of the century however, largo fishing had become completely unprofitable because of the continually rising costs of long periods at sea and the depletion of stocks at Reed Bank.

Technological change in fishing gear has proceeded, with continual improvements to enable greater catch for less effort. Over time, net mesh sizes have become smaller and, in particular, fish-finder technologies introduced during the fusilier-fishing era in the early 1990s were seen as a major innovation. Many of the technological innovations to fishing gear, such as the modifications to existing gillnets, have originated from the Visayan Islands. Knowledge of them was transported to the Calamianes through migration or when a local resident fished for a time elsewhere. The fact that Visayan fishers have developed many fishing techniques has been reported elsewhere in Palawan (Eder 2008: 86) and the Philippines (Spoehr 1980).

The central seasonal distinction for fishers in Esperanza is between the northeast monsoon (amihan), which runs from approximately October to early May, and the southwest monsoon (*habagat*), which runs from approximately late May to September. Amihan is characterised by dry, consistently strong winds that make fishing in small vessels very difficult and the fusiliers commonly targeted by captains become difficult to catch during this period. Hence amihan is the season of hardship in Esperanza, characterised by freshwater shortages, requests for credit from poorer families, difficulties navigating the seas and less fish at the market. Habagat on the other hand, is characterised by the milder southwesterly winds that bring consistent rain and more gentle seas off the eastern coast of Palawan. This is the season of relative wealth in Esperanza, when families are busy with the catching, transport and marketing of fish. Perhaps the simplest characterisation of the differences between the two seasons came from one fisherman: 'In *amihan*, we drink Emperador [cheap brandy]. In *habagat*, we drink beer!'[6]

While there is no specific 'market day' by which the fishers have to regulate their activities, lunar cycles play a strong role in determining when they are able to fish. The full moon period is inevitably bad for most fishers, and at these times, squid dominate the market. Tidal patterns are also significant in regulating the times of departure for fishing vessels.

While a key characteristic of fishing in Esperanza is its flexibility and diversity, it is possible to characterise four primary fisheries: net fishing (targeting rabbitfish, see Plate 2-6), fusilier fishing (targeting fusiliers from the Family Caesionidae, see Plate 2-7), fresh grouper (targeting leopard coral grouper) and live reef fishing (targeting and then keeping alive leopard coral grouper (Plates 2-2, 2-3).[7] Net fishing and live reef fishing are both 'small-scale' in that they are not characterised as commercial fisheries under Philippine law, which stipulates that a commercial vessel is one which weighs more than three gross tons. These types of fishing are conducted on smaller pump boats with less powerful engines. In contrast, fusilier fishing and fresh grouper fishing are both commercial activities. While the season of amihan is generally less favourable for all types of fishing, the only strictly seasonal fishery is the fusilier fishery, which only runs during habagat. Almost all of the fishers in Esperanza are engaged to varying degrees with one or more of these four major fisheries.

Plate 2-6: Rabbitfish caught by net fishers.

---

6   Beer being considerably more expensive than any of the local spirits.
7   Chapter 3 describes the economics and other issues related to these four fisheries in greater detail.

Plate 2-7: Red-bellied fusilier caught by fusilier fishers.

Source: Photo courtesy of Al Linsangan.

Women occasionally fish with their husbands on small-scale vessels, but are not involved in commercial fishing. They do, however, have a central role in the marketing of fish and control over household finances (see Eder 2006). For many fishing families that rely primarily on net fishing, for example, it is the wife who will often bring the fish to the market in Coron town and negotiate with the buyers at the market. Similarly, in the commercial fusilier and fresh grouper fisheries, it is frequently the wife of the boat owner who will take charge of crew payments and trip financing.

A wide range of other fishing techniques and gear are used in Esperanza. Another common method used by fishers unable to afford pump boats is simple hook-and-line, using a basic one-person boat and paddle power. These fishers target groupers and other first-class fish. During early mornings and evenings, some couples will wade along the shallows with a lamp and a net, capturing shrimp. At other times when the tide is out, women and children will comb the tidal flats, gleaning for shellfish (Plate 2-8). Some fishers maintain small fish cages to keep live groupers for grow-out (Plate 2-9). During the full moon period, when fish are wary of being caught, many fishers will use squid jigs to catch squid instead. Other techniques and gear used include spears, long-lines and crab nets. Still other residents form connections with the commercial 'baby' purse seine boats (*pangulong*) that sometimes stop outside the village, exchanging products for fish and then subsequently selling the fish.

Plate 2-8: Gleaning for shellfish at low tide.

Plate 2-9: Building a fish cage.

An important point to stress about these patterns of fishing therefore is their flexibility. Many fishers, for example, will work with hook-and-line on a commercial fusilier boat during habagat, but then work on a net fishing boat during amihan. Others will work for some fusilier trips only, and then make trips with live reef fishing boats for the rest of the year. Fishers will make their decisions about what gear to use, where to fish and which species to target based on a range of fluctuating economic, personal and environmental factors. It is of little use therefore to give statistics on the number of fishers who use particular gears and methods, as the members of the different fisheries are so interchangeable. Instead, households aim for occupational mobility within the fisheries sector, constantly adapting to changing economic and weather conditions. I have aimed here to merely provide an historical perspective and an introductory sketch to the fisheries of Esperanza.

Conclusion

I have shown in the preceding sections how natural resources are vital to the economic and everyday lives of Palawan at a provincial, municipal/regional, and local level. Conservation, tourism and commercial resource extraction are the primary ways in which groups of people are attempting to use or manage these resources. In the Calamianes, dive tourism, commercial fishing for export and marine conservation are the main modes of resource use. In Esperanza, everyday life is still dominated by fishing.

All three modes of resource use are at the same time adopted as strategies by governments at multiple levels in Palawan. All three modes are closely related, complementing and contradicting each other, and each can be seen as a particular way in which visions of the 'frontier' are enacted. As with my treatment of the poverty-environment relationship in this book, the focus in this chapter has been to adopt a perspective informed by post-structural political ecology to demonstrate not only the extent to which these three modes of resource use occur, but also how they are understood, represented and contested by various actors. Governments see tourism as a pathway to economic growth and development, and yet tourism relies on and is closely linked to the presence of conservation. Commercial resource extraction, particularly fishing, is still seen as essential and worthy of investment. Different administrative levels of the state, tourism operators, conservationists, businessmen and local residents all have different relationships with conservation, tourism and resource use. How the particular consequences and effects of these relationships play out in different situations is what I explore in the following chapters.

3. Economic, Class and Status Relations in Esperanza

The purpose of this chapter is twofold. Firstly, it will provide a detailed background to the sorts of social and economic relations that residents of Esperanza are frequently involved in, focusing in particular on relations in the fishing sector. Secondly, it will show how these relations can be characterised with reference to the idea of poorer people entering into reciprocal relationships with, and making claims on, those with more resources. Asking other residents to help pay for school fees or medicine; gaining access to a financier to fund equipment and individual fishing trips; moving to an alternative fishery; getting a position as a crew member on a boat; all these practices tend to require reciprocal relationships between people of unequal status. While there is great diversity in the nature and consequences of these relationships, this chapter will describe how these sorts of relationships play out in Esperanza, aiming to give a sense of the everyday economics and politics of life in the sitio. I focus in particular on such relationships within the context of fishing.

In order to situate my discussion, this chapter begins with a review of some of the voluminous literature that deals with social and economic relations in the Philippines. After a brief discussion of the ways in which I have categorised different class and status groups in Esperanza, I examine firstly class and social relationships within various everyday contexts. Drawing on these themes, I then consider how personalised economic relationships play out in the four primary fisheries of Esperanza. The bulk of the chapter examines in detail the economics underlying the four primary fisheries of Esperanza in order to understand broader class and status relations: how people finance their fishing trips, how much money is involved in fishing trips, how profits are shared, and how and where the fish are sold. I do not intend to make a neat argument about how the social and economic relations of the fisheries clearly mirror those in broader everyday life. Each of the fisheries I describe holds a range of different relationships. Instead, this section will show how such economic patterns are broadly characterised by personalised reciprocal relationships, both among different classes within Esperanza, and between different classes in Esperanza, Coron, Manila and elsewhere.

Approaches to Economic, Class and Status Relations in the Philippines

There is a vast literature that deals with social and economic relations in the Philippines. Here I have broadly characterised three distinct, yet related approaches—one focusing explicitly on patron-client relationships and class relations, a second focusing on some of the cultural concepts in the Philippines associated with such relationships, and a third focusing on patterns of economic personalisation. Each approach has its own set of debates, but it is not my intention to engage with the particularities of such debates here. Instead, my purpose in analysing this literature is simply to contextualise my analysis. I merely aim to give a flavour of how people of unequal status initiate and maintain reciprocal personalised relationships in the context of a coastal village such as Esperanza.

Class Relations and the Patron-Client Relationship

Research on class relations in the rural Philippines has been dominated by studies of patron-client relationships between landlords and tenants (Scott and Kerkvliet 1977; Fegan 1982; Wolters 1983). As Scott and Kerkvliet (1977: 440) define it:

> [a] patron-client link is an exchange relationship or instrumental friendship between two individuals of different status in which the patron uses his own influence and resources to provide for the protection and material welfare of his lower status client and his family who, for his part, reciprocates by offering general support and assistance, including personal services, to the patron.

Frequently, a theme of this literature is to argue that the exchange relationships between patrons and clients have been historically transformed because of various factors, leading to a worse position for the 'landless labourers' (Fegan 1982) and other 'clients'. Scott and Kerkvliet argue with reference to Southeast Asia generally, for example, that '[r]ural class relations that had once rested, in part, on consent became, under the forces of commercialization and colonial government, increasingly characterized by coercion and exploitation' (Scott and Kerkvliet 1977: 455).

Other writers have sought to use the patron-client concept to explain political life in the Philippines. Hollnsteiner (1963) and Landé (1965) showed how political patrons would offer goods and services to particular regions in exchange for voter support, and this model has subsequently remained a popular one among observers of Philippine politics (Kerkvliet 1995: 401; Kasuya and Quimpo 2010).

In this model, various 'factions' form a pyramid, with the highest political patron at the apex, and their supporters massed below. Thus political and class relations become characterised as 'symbiosis' (Agpalo 1972, cited in Kerkvliet 1990: 243), and the links between poor people and rich people are emphasised and characterised as 'mutual aid' (Landé 1965, cited in Kerkvliet 1990: 243).

Various writers critical in particular of what they saw as an overly harmonious view of Philippine class relations, and as ignoring the potential for conflict and tension between different classes (Ileto 1979; Rafael 1988; Kerkvliet 1990; Cannell 1999; Roces 2001) subsequently critiqued analyses focusing on patron-client linkages. Kerkvliet (1990) for example, focuses on the resistance of poor peasants to richer landowners in his study of 'everyday politics' in San Ricardo, Central Luzon. He argues cogently that while patron-client ties are undoubtedly still important, this does not mean that conflict does not exist. Instead, he argues more generally that '[a] central dynamic of everyday politics in San Ricardo society is people trying to make claims on each other and on a range of resources according to their relationships to those superordinate or subordinate to themselves and in terms of their interests and values' (ibid.: 14). With regard to the patron-client, factional model of political life, Kerkvliet has also argued that it 'leaves little or no room for other values and ideas, other bases for cleavage and struggle, other grounds for organizing and cooperating' (Kerkvliet 1995: 404).

I intend to argue that the forms of patron-client linkages that exist in Esperanza are somewhat different to the classic landowner-tenant relationships described in Luzon. Instead, like Kerkvliet (1990: 14), I aim to take a broader view, and simply suggest that poorer people make a variety of claims on richer people that are expressed in a range of ways. A key aspect of everyday life in Esperanza that I want to emphasise is how poorer fishers appeal to and make claims on other more well-off fishers, households and financiers. These appeals and claims are frequently expressed through particular cultural idioms of the Philippines.

Debt and Shame

The second, related approach to exchange that I introduce here is therefore one that has focused on some of the cultural ideas underlying exchange and reciprocity (Kaut 1961; Bulatao 1964; Hollnsteiner 1970; Lynch and De Guzman II 1970). In an influential article, Hollnsteiner (1970) distinguished between three types of reciprocity in the lowland Philippines: contractual, quasi-contractual and 'debt of the inside/heart' (*utang na loob*) reciprocity. Contractual reciprocity, according to Hollnsteiner, is where the reciprocal acts are 'equivalent': the amount and form of repayment is clearly established, and there is little, if any, emotional aspects to the exchange. In quasi-contractual reciprocity, by contrast, the terms of exchange are implicit and generally culturally defined. Hollnsteiner gives the example of a cooperative workday (*bayanihan*), where a man would ask all his friends and neighbours to help build a new fishing raft. In any future cooperative workday of those friends and neighbours, the man would then be expected to help them in return. The key feature of quasi-contractual reciprocity, according to Hollnsteiner, is where 'no clear statement of obligation has been made by either party, yet the necessity to repay in kind when the opportunity to do so arises is mandatory' (ibid.: 69).

The third type of reciprocity distinguished by Hollnsteiner (1970), utang na loob refers to an enduring, personal debt that cannot be quantified, and is often associated between people of unequal status. Some versions of this type of reciprocity can never be fully repaid; a typical example in the Philippines is the debt a child owes to his/her parents. Other examples cited by Hollnsteiner that create utang na loob include sending a friend or relative's child through school, facilitating entry into a hospital for someone who is sick, or getting a job for someone. Closely related to utang na loob is the idea of *hiya*—usually translated as shame, shyness or embarrassment. As Bulatao defined it in the early 1960s, hiya is 'a painful emotion arising from a relationship with an authority figure or with society, inhibiting self-assertion in a situation perceived as dangerous to one's ego' (Bulatao 1964: 428). Hollnsteiner (1970: 71) similarly translates it as 'a sense of social propriety'; if utang na loob cannot be repaid, for example, it is considered a great shame and the individual will feel hiya. Indeed, merely asking for any sort of debt will normally generate hiya and must be managed diplomatically. As Hollnsteiner puts it, '*hiya* is not necessarily accompanied by *utang na loob*, but *utang na loob* is always reinforced by *hiya*' (ibid.: 82).[1]

While Hollnsteiner and other members of the 'Ateneo School' (Lynch and De Guzman II 1970) have been criticised for their tendency towards structural functionalism (Cannell 1999: 8–9), and for their inadequate treatment of how these concepts are tied up in class relations (Pinches 1991), I would argue that their work remains useful as a general introduction to how these sorts of principles can operate in the Philippines. In Esperanza, the cultural norms and values surrounding reciprocal exchange are present to a strong degree.

As I shall further show in Chapter 4, these sorts of reciprocal relationships between people of unequal power in the Philippines are often enacted through the idiom of pity and a 'basic rights discourse' (Kerkvliet 1990; Cannell 1999).

---

1 Hollnsteiner makes the point that these forms of reciprocity can be seen in the personalisation of public institutions: 'The concept of impersonal service is not deeply ingrained in the bureaucracy or the general public; gift giving and receiving for service rendered is common…. How can the fulfilment of one's social obligations, brought about through *utang na loob*, be anything but good, reasons the average Filipino' (Hollnsteiner 1970: 76). Mulder (1997: 132–3) and Pertierra (2002: 88–90) both make similar points, and as I will show in Chapter 7, this has significant implications for how rural people in the Calamianes perceive the role of government.

Poorer people, according to Cannell (1999), 'ask for pity'—not necessarily by presenting themselves in a pathetic way, as the English term may suggest, but by appealing to a 'right to survive' (Blanc-Szanton 1972), and to an ethic of fairness for the poor. I want to show how in Esperanza, many of the sorts of reciprocal relationships involved in fishing and in everyday life are mediated through the sorts of concepts I have introduced in this discussion.

Personalised Economic Relationships

The third approach to economic and social relationships that needs to be explored is one that has focused on patterns of economic personalisation. Economic anthropologists have long documented the existence of markets where traders habitually sell to a regular partner. Forms of these personalised economic relationships were documented originally in Haiti by Mintz (1961) as *pratik*, and in the Philippines by Davis (1973) as *suki*. As Davis describes, the term *suki* refers to both the customer and the seller, and 'nearly every seller has "supplier *suki*" from whom he buys and "customer *suki*" to whom he sells. The usage of the term is reciprocal, each partner referring to the other as "my *suki*"' (ibid.: 217). The extent of the relationship may vary in intensity. Sometimes, suki may simply refer to a favoured or regular customer; in other instances, favoured prices and business and personal credit may form an integral part of the relationship. Davis argues that there are significant advantages for both buyers and sellers in this relationship. Primarily this is the security of ensured outlets for sellers, and the presence of continuous supply for buyers (ibid.: 226–7).

Similarly, Dannhaeuser (1983: 52) argues that suki relations increase security:

> in an insecure economic environment. Buyers seek reliable suppliers who give credit, preferred selection of new or scarce goods, and reasonable prices without the need to haggle incessantly. Suppliers search for regular outlets who won't constantly drive a hard bargain and who can be trusted in matters of credit. *Suki*, in short, is an informal way to stabilize commercial relations.

Essentially, the idea is that producers and buyers forego the opportunity of potentially high short-term benefits for the stability of a regular exchange relationship based on a level of trust.

Firth (1966: 126–84) and Szanton (1971: 53–4) both describe a particular form of personalised exchange among fishing communities of Southeast Asia, whereby capital is essentially traded for labour. In Szanton's ethnography based in Panay in the Philippines, fish buyers would extend credit to fishers in the form of money or equipment. The fishers would then catch the fish and sell them to the buyer at a price slightly lower than market value. Szanton shows that such relationships favoured the dealer who was able to secure a very high return on the invested capital, when compared to the fisher's return on his labour (ibid.: 54–6). Despite this, he argues that '[t]he dealer's ready supply of cash and credit for fishing operations and consumption items makes the relationship advantageous for the fisherman as well' (ibid.: 57).[2]

Other writers have since critiqued this positive perspective of personalised relationships. Russell (1987: 139) argues that such models ignore power relations inherent in personalised exchange: '[t]hey give little attention to the wider institutional constraints that limit the choice of contractual arrangements nor to the potential power of personal relations to enhance trading imperfections'. Russell demonstrates that in the Philippine highlands, vegetable farmers were effectively forced to participate in personalised exchange relationships simply if they were to enter the market at all; 'the great degree to which farmers are indebted to a middleman prior to the sale of the harvest gives middlemen substantial power' (ibid.: 150). Subsequently, many farmers had no choice but to accept exploitative trading conditions such as poor prices. Thus, Russell argues that the basis of the entire personalised exchange is the initial inequalities between the trading partners. As Alexander and Alexander put it, citing Russell's analysis of the vegetable market in the Philippines, and Acheson's (1985) description of the relations between fishermen and dealers in the USA, 'in both cases producers can choose their partners; but in neither case can they refuse to dance' (Alexander and Alexander 1991: 502).[3]

Based on fieldwork in Panay, Rutten argues that certain questions firstly need to be answered before it can be concluded that credit is 'good' or 'bad'. The level of demand for products, the degree of choice in accepting credit, whether or not it is also possible to sell in an open market and the level of competition among buyers are all factors that need to be taken into account in the analysis of credit relations (Rutten 1991: 116). Bearing in mind the accounts of Russell (1987) and the Alexanders (1991) that demonstrate how producers in many cases cannot afford not to enter into personalised relationships, I have included Rutten's argument at the end because I suggest that the relationships I describe in this chapter are highly variable in terms of their economic effects. Here I do not intend to engage in a detailed discussion of whether such relationships are

---

2   Plattner argues that both of these sorts of personalised economic relationships (suki and extending credit) are 'equilibrating'. This means they are 'generalized, reciprocal, open-ended, and long-run' relationships where each partner has trust that 'an imbalance in an exchange will be made up in future exchanges' (Plattner 1983: 849). Plattner stresses the economic rationality inherent in such relationships, arguing that they are a rational economic response to the weaknesses inherent in rural agrarian markets: '[v]ariable-quality goods, cheap labour and scarce capital, and insufficient transportation and communication systems mean that traders need to rely on equilibrating relationships to stabilize their businesses' (ibid.: 851).

3   Similarly, analysing the central vegetable market in Cebu City in the Philippines, Hendriks argues that '[u]nequal bargaining positions exist where farmers and small traders depend on credit contracts' (Hendriks 1994: 6).

exploitative or beneficial to all parties; instead, I want to merely introduce some of these themes of personalisation, and show how they figure in the everyday lives of coastal residents in Esperanza.

Everyday Social and Economic Relations in Esperanza

Categorisation of Classes in the Philippines and Esperanza

Firstly, a quick discussion about categorisation is necessary. Eder (1999) makes the important point that much of the literature on class and social relations in the Philippines has been dominated by studies on the rice-growing region of Central Luzon (for example Hollnsteiner 1963; Fegan 1982; Kerkvliet 1990). As he points out, provinces in this region, such as Bulacan and Nueva Ecija, have been 'long characterized by monocrop agriculture and highly asymmetrical patterns of land ownership' (Eder 1999: 73). As I outlined in Chapter 2 however, my research is concerned with predominantly fishing households where agricultural, landlord-tenant relationships are not present. How class relations play out in such coastal contexts is something that has received less attention in the academic literature (but see Russell 1997; Eder 2008). Indeed, generally throughout Palawan, including in the agricultural sector, such calcified, seemingly eternal class distinctions are not as present, where smallholder operations are more the norm (Eder 2008: 48).

Classifying the households of Esperanza into differential classes, therefore, is not as clear-cut as the classic Marxist division between wage labourers and capitalists, or landlords and tenants. Other writers have also struggled with such distinctions, even in the more strongly differentiated rice-growing regions of Central Luzon. Kerkvliet, for example, found that household class differences were often crosscut by differences related to status, and that households frequently contained individuals belonging to different classes (Kerkvliet 1990: 77). Notwithstanding these complications, he ended up dividing the community he studied into four classes: workers, peasants, small business owners and capitalists. Within these classes involved further subdivisions of workers into three categories, and peasants into five subdivisions. In his extensive study of economic change in mainland Palawan however, Eder (1999) found that it would be too problematic to use the term 'class' to represent different groups within the community he studied. He stated that:

[t]he multiplicity of on- and off-farm economic activities in the community is such that many residents enter into a variety of different economic relationships at the same time, making impossible the demarcation and juxtaposition of clear-cut collectivities of households standing in some type of stable relationship with one another, or with the means of production, over time (ibid.: 6–7).

Although class distinctions in Palawan are frequently not as clear-cut as in other regions of the Philippines, as I pointed out in Chapter 2, however, because of the lack of land tenure and agricultural opportunities, Esperanza maintains a greater marine orientation than many other coastal areas of Palawan and the Philippines. This means that a greater proportion of households are concerned primarily with fishing, and so it is possible to identify different fishing groups or classes within the community. While class distinctions are by no means rigid or clear-cut, I argue it is possible to roughly differentiate three broad groups of fishers: crew members; small pump boat owners; and commercial boat owners. The flexibility of fishing activities described in Chapter 2 makes it difficult to be any more specific about class, and makes the distinctions I have made only approximate ones. Bong, for example, was an engine-operator (*mekanista*) on a fusilier-fishing boat during habagat, and captained his own net fishing boat for the rest of the year. I have simplified the situation therefore by including fishers like him, who belonged to several different groups, in the 'highest' rank that they had. The typical features of these groups are summarised in Table 3-1.

Table 3-1: Socio-economic features of fisher groups in Esperanza.

Group of fisher	Monthly income (₱)	Primary economic activity	Household circumstances
Crew member	500–5000	Has to continually secure a place as a crew member, often in various net and commercial fisheries. Sometimes owns small canoe for simple hook-and-line fishing.	No electricity; no gas for cooking; no appliances; bush material houses.
Pump boat owner	3–10 000	Owns small pump boat for net fishing or live grouper fishing; usually captain of boat but will sometimes hire boat out.	Often electricity and gas; some limited appliances.
Commercial boat owner	Above 10 000	Owns commercial fusilier or fresh grouper boat; often captain of the boat as well but will sometimes hire boat out.	Electricity; multiple appliances; concrete houses.

Source: Author's calculations, based on census data, household survey and interviews, 2006.

Generally, membership of these three different groups in Esperanza correlates with the level of wealth. Crew members are the poorest fishers, and as one of them told me, they 'have to fish for every single *centavo* [cent]' they earned.

These fishers often have a particularly high level of flexibility in the fisheries they practice, moving between net fishing, fusilier fishing, grouper fishing, and supplementary fishing activities (for example, gleaning or shrimp fishing), depending on seasonal and personal circumstances. I have also included in this category fishers who are not necessarily always crew members on net fishing or commercial boats, but who usually paddle a canoe out to the reef or sea beyond and fish with hook-and-line. Monthly incomes for these fishers range significantly depending on the season, the particular fishery they are involved in and variability in fish catch (which is dependent on a whole range of other factors as well), but generally these fishers earn well below ₱5000 and sometimes, such as during amihan, they will earn little or nothing at all. Some of these fishers will often struggle to make ends meet on a daily basis, especially if they have a family to support. Don, a fisher who usually practiced net fishing but sometimes joined on some fusilier-fishing trips during habagat, was typical of this category of fishers. Houses of fishers like Don are often primarily made of bush materials, they frequently have no access to electricity, and the main dish accompanying rice (*ulam*) will usually be based around fish and vegetables (as opposed to meat such as chicken or pork).

The mid-level economic category is characterised by fishers who operate a small pump boat collecting fish for the live fish trade and net fishing. As with the crew members, incomes are highly variable among this group, ranging from around ₱3000 (during amihan) to ₱10 000 (during habagat). Households of these middle-income fishers can afford electricity and some simple appliances like a stereo and gas stove for cooking. In Esperanza, the most economically successful fishers in terms of wealth and levels of capital invested in the fishing fleet are the owners of the commercial fishing boats—the fusilier and fresh grouper fisheries. These households have concrete houses, sometimes toilets with septic tanks, and appliances such as stereos and televisions. While those in the fusilier fishery were struggling financially at the time of fieldwork and often moved from debt-to-debt, by virtue of owning a large, commercial boat, I have classified them as belonging to a higher economic status.

Also relevant for the purposes of this chapter are the traders, financiers and other elites who live outside of Esperanza. This 'category' is considerably more mixed than the categories I have defined within Esperanza. For instance, the traders who buy rabbitfish from the net fishers in Esperanza and sell it at the Coron town market are very different to the live fish traders, many of whom held significant political influence at the time of my fieldwork research. So instead of summarising their typical features I will elaborate on the circumstances of these particular actors with reference to the fishery with which they are involved; discussions of which follow.

Following Kerkvliet (1990: 61), I found social status in Esperanza to relate primarily to wealth, or simply the standard of living within a household. Although the level of 'respect' individual households have varies according to their behaviour such as whether they treat poorer people well (see Kerkvliet 1990: 61–2 for a similar understanding), status here is understood in the narrow sense of wealth. While social hierarchies between these different groups in Esperanza are not as stark or pronounced as those documented by Kerkvliet (1990) and other writers in Central Luzon, similar general patterns of reciprocity exist.

Class and Social Relations in Esperanza

The Marquez family was regarded by the sitio community as one of the most high status families in Esperanza, and was demonstrably one of the richest. They were typical of a higher status family in that they owned a commercial boat(s), and their material standard of living was very high when compared to other families in Esperanza. They owned a television and a DVD player; they ate meat with many of their meals; their house was made of concrete and they had a toilet with a septic tank; and they could afford to send their children to private schools in Coron town. Manny and Grace arrived in 1968 from northern Samar with their daughter Anna. As such, they were one of the first families to settle in Esperanza. While Manny and Grace subsequently moved along the coast to a neighbouring sitio, Anna married Carlos Marquez, who had migrated from Romblon in the early-1970s. Carlos began working on basnig boats in Coron Bay during the 1970s, and by 1981, the family was relatively successful and owned their own basnig boat. Their fishing business continued to expand with largo fishing during the 1980s, fusilier fishing in the 1990s, and by 2006, the family had the biggest fishing business in Esperanza. Carlos Marquez based himself in Puerto Princesa with his fresh grouper boat, his son-in-law managed a large fish cage of several hundred leopard coral grouper in Esperanza, his brother-in-law captained a fusilier boat out of Esperanza, and his wife and daughter managed the finances (see Figure 3-1).

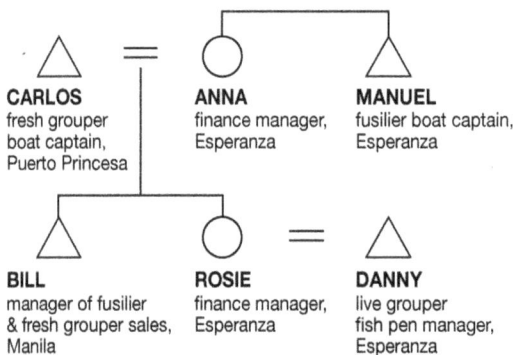

CARLOS
fresh grouper
boat captain,
Puerto Princesa

ANNA
finance manager,
Esperanza

MANUEL
fusilier boat captain,
Esperanza

BILL
manager of fusilier
& fresh grouper sales,
Manila

ROSIE
finance manager,
Esperanza

DANNY
live grouper
fish pen manager,
Esperanza

62 **Figure 3-1: The Marquez family and their business relationships.**

I describe later how their strong financial position led to their financing the fishing trips of other members of the community on occasion, but first I want to focus on some of the ways in which their relations with other community members functioned in other contexts. Frequently, during the dry, difficult season of amihan, children from poorer households in Esperanza would regularly come knocking at the door of the Marquez family house. Rice was a common request, as was money for school fees. Being able to afford to send all the children in a family to school is often a challenge for parents in Esperanza. This is especially so if such a family engages in what is locally known as 'family planting', an ironic pun on 'family planning' and a practice that produces numerous children. One morning for example, a child from a poorer family came to the window of Anna's house while I was having coffee with her. Although she was already eleven-years old, she was still hopeful to begin elementary school, something that had been postponed because her family had not been able to afford her school fees previously. The child explained her family situation, which relied mostly on the income earned by her father, a poorer fisher who used hook-and-line on a paddle-powered canoe. She made a great deal of the fact that her family would not be able to send her to school without the generous help of a benevolent woman such as Anna.

Key here are concepts that I introduced in the earlier discussion of exchange and reciprocity. Using a child as a 'go-between' reduced the level of shame (hiya) and awkwardness by those who needed the assistance. Through the child, the parents were trying to enter into a reciprocal relationship with the Marquez family, presenting themselves as 'pitiful' (Cannell 1999) in order to obtain assistance that would be repaid in some manner in the future. By helping a poor family with such an important thing as sending a child through school, the Marquez family was probably accruing a form of utang na loob, and as such, the form and quantity of the repayment was uncertain. And, while the way in which Hollnsteiner strictly demarcates various forms of reciprocity into various categories inevitably simplifies the situation somewhat, this sort of debt between the child and Anna was clearly not 'contractual'. When I asked Anna Marquez about her expectations for repayment, she told me that while it would be nice if the family of the child repaid the school fees in full, in reality she expected that this would be difficult to achieve. She said that the family would probably contribute in other ways, such as bringing around dishes or fish that they caught. As a foreign resident, I also found myself the subject of numerous requests for help to do with medical costs, rent and food in Coron town and Esperanza. Frequently, residents asked me to 'share [my] blessings'; appealing to a sense that those in a more well-off position should help those less fortunate.[4]

For the Marquez family, fulfilling requests such as these was usually obliged. For them, it was important to maintain 'an ability to get along with people'

---

4   As Hollnsteiner describes it, 'the social system requires that those who have more should share their bounty with those who have less' (Hollnsteiner 1970: 67).

(*pakikisama*); literally meaning 'togetherness' with the rest of the community. They were conscious of the fact that they were well off compared to others in Esperanza, and to them it was very important that they be seen as good-hearted and generous. Practically, too, they said that it was important to have 'happy workers'. As they employed many members of the community on their commercial boats as crew, it was important to them that the workers were satisfied. During the process of unpacking the commercial fusilier and fresh grouper boats, for example, many people would come along to assist. These would often be people with no formal connection to the crew or the family, yet they would put in great effort unloading the boxes of fish or repacking the fish in ice. For such work, Anna would then be obliged to pay them something (usually something simple such as snacks) for their contribution.

Their reputation for generosity to poorer members of the community was strong. Bong, for example, was a fisher who worked as an engineer on one of the Marquez family commercial fusilier boats. He stated that he would regularly go to them during amihan, rather than other more well-off families, because of their kindness. A short anecdote illustrates the strength of the positive perceptions of the Marquezes. Maricel, a poorer woman from the community, had recently finished her contract job as a domestic helper in Coron town. Her husband, Don, was a net fisher who was often unable to make enough money to support their family, which included eight children. Carlos and Anna decided to employ Maricel as a domestic helper in their household in order to help her. As Anna put it, 'I don't need a domestic helper here. I can look after my own place. We decided to employ Maricel because of *makatao* (humanity)'. Here, again, Maricel was deserving of pity (Cannell 1999). Several months into this relationship however, a misunderstanding over payment led Maricel to cease working for Anna and accusing the Marquezes of underpayment. All corners of the community were awash with gossip over the next few days; however, the consensus among other households was that Maricel must have misunderstood the payment method and that the Marquezes should not be to blame for the angry fallout.

For the poorer people in Esperanza, assistance from well-off families like the Marquezes was gratefully received. Recipients of such largesse would repay favours in different ways. Fine specimens of freshly caught fish would be given to the Marquez family or sold to them for a cheaper price. Members of poorer households who were in debt would provide labour around the house such as laundry work, or do extra work on the boats. Frequently, such relationships between wealthier and poorer households are forged through connections of kin or work. Raul, for example, was an elderly man who eventually found fishing on the Marquez commercial boats too difficult because of his age. His children had moved away and were no longer in contact and with his frailty; his circumstances rendered him with few livelihood options. The Marquezes gave him work to guard their fish pen during the night. Again, this was seen as a non-contractual debt: giving work to somebody who had few or no means of attaining it otherwise was not able to be repaid in a simple or contractual way. For Raul, 'I have *utang* [debt] towards them because they helped me when I really needed some work. I try to pay them back by doing a good job, making sure nobody takes any fish from their pen. Sometimes I go fishing near the fish pen and give them my catch in the morning'.

Similar relationships exist between households in Esperanza, and other households in Coron town. The Marquez family, for instance, maintained links with the Alejo family in town. The Alejo family had been one of the large basnig boat operators during the 1970s, and Carlos Marquez had originally started fishing in Coron for Mr Alejo Snr. Since this time however, the Alejo family have moved away from fishing and have achieved relative success in politics with Mr Alejo Jr acting as member of the municipal council during the 1990s. The Marquez family had not had a formal business or fishing relationship with the Alejos since the 1980s; instead, they maintained an informal friendship and contact. When they required a loan for capital or other expenses, they would sometimes call on the Alejos for financial support. In return, the Marquez family ensured the Alejos received political support in Esperanza. When the mother of Mr Alejo Jr died in 2006, the Marquez family offered various assistance during the mourning period, such as contributing gifts and helping to organise mourning activities.

None of this is to say that all relations between poorer and richer households in Esperanza are necessarily harmonious and free of conflict. Some richer families were cast as stingy, and there was certainly evidence of some of the tensions that Kerkvliet (1990) focused on in his analysis of 'everyday politics'. The point I am making, however, is that personalised modes of reciprocity form the basis of much of the social relations between households of different status in Esperanza, and that such reciprocal relationships are often marked by certain cultural idioms.

The Social and Economic Relations of Fishing in Esperanza

Personalised economic relations exist in multiple arenas within the fisheries of Esperanza. Pumpboat and commercial boat owners and their respective crews, buyers and sellers of fish, financiers and owners of boats all practice variations of personalised economic relationships. My approach here is not to describe them in great depth, although some of the typical features are summarised in Table 3-2 and Map 3-1. Instead, I aim to outline the basic economic patterns of the fisheries in Esperanza and how they are intertwined with the themes that I have introduced in this chapter so far: poorer people making claims on those with more resources through particular cultural idioms, and the formation of economically personalised relationships. In each fishery, I will demonstrate how the class relations that I have introduced in this chapter are expressed in various relationships related to fishing. It should be emphasised though that I do not intend to imply any sort of neat argument or equivalence between all of these relationships—class and status relations are similar to, but do not necessarily replicate, the other forms of relations I have already described. At the risk of thematic overlap, I have organised discussion around the four primary fisheries: net fishing; fusilier fishing; the fresh grouper fishery; and the live grouper fishery.

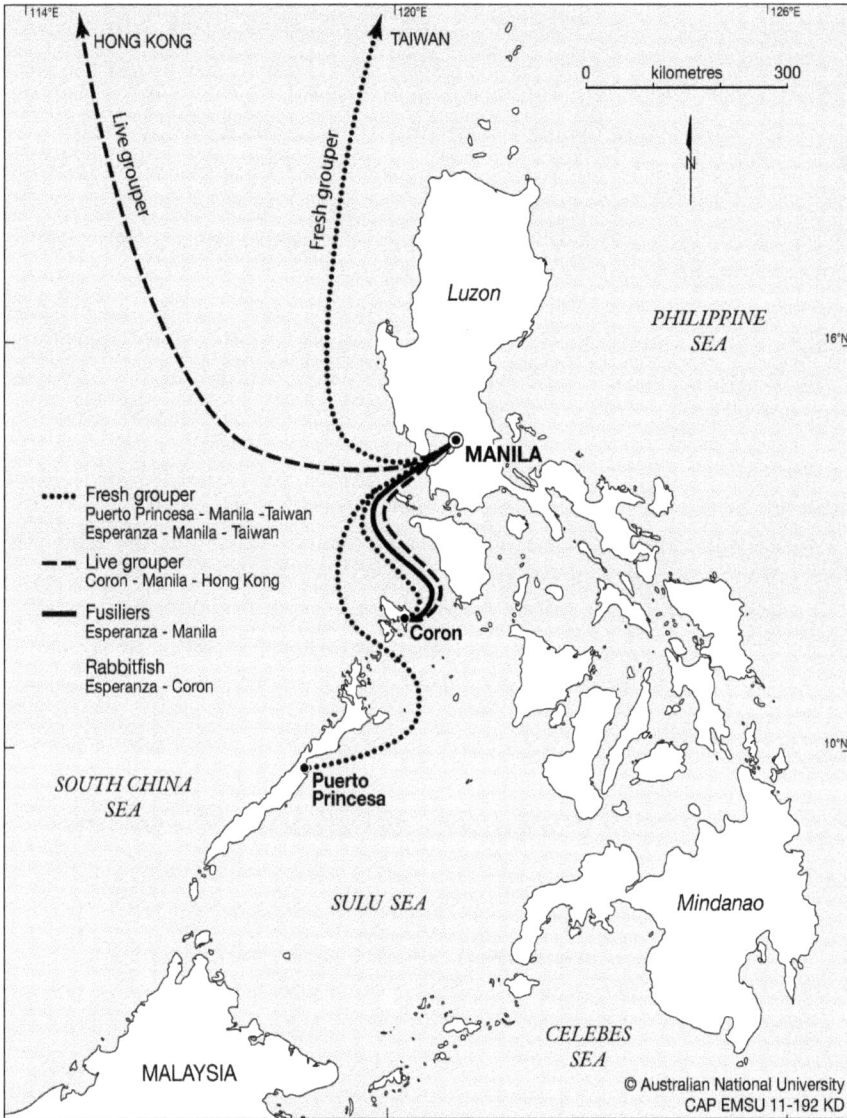

Map 3-1: Map indicating fish export destinations.

Source: Cartography ANU.

Table 3-2: Economic characteristics of primary fisheries in Esperanza.

Fishery	Crew (number)	Trip Expenses (₱)	Length of trip (days)
Net fishing	2	800–1200	1–4
Live grouper	4	5–6000	4–10
Fusiliers	20–30	70–80 000	10–14
Fresh grouper	25–30	200 000	10–14

Source: Author's calculations, 2006 data.

Net Fishing

While there are many variations in specific gear and technique, the type of net-fishing mostly practiced in Esperanza involves the use of a plunger and gillnet (Plates 3-1, 3-2). It is conducted from a double-outrigger pump boat, around five metres long, powered by outboard motors ranging from 6–16 hp. There are normally two crew members on the boat. During a fishing expedition, one fisher will pilot the boat while the other sets down the gillnet along the seagrass bed. With the net in place, the fishers thrust a long wooden pole plunger into the water from the boat while it traverses the length of the net, driving the fish into the net. Target species include various types of rabbitfish (Family Siganidae). Fishers in this area adopted this particular form of gillnetting in 1989, when a local fisher returned from several years fishing in nearby Batangas Province. Trips can range in length from just a couple of hours fishing along the village shoreline to three or four days. During longer trips, fishing is usually conducted at more remote locations that have more fish. There is no strict season for net fishing; however from June to February there are plenty of fish to be caught but the price is lower (₱25–40 per kg). From March to May the market price increases (₱50 per kg) due to smaller catch sizes. The fish are usually sold at the local market of Coron, and are sometimes dried (see Plate 3-3) and later bought by roving dealers or middlemen based in nearby provinces such as Occidental Mindoro.

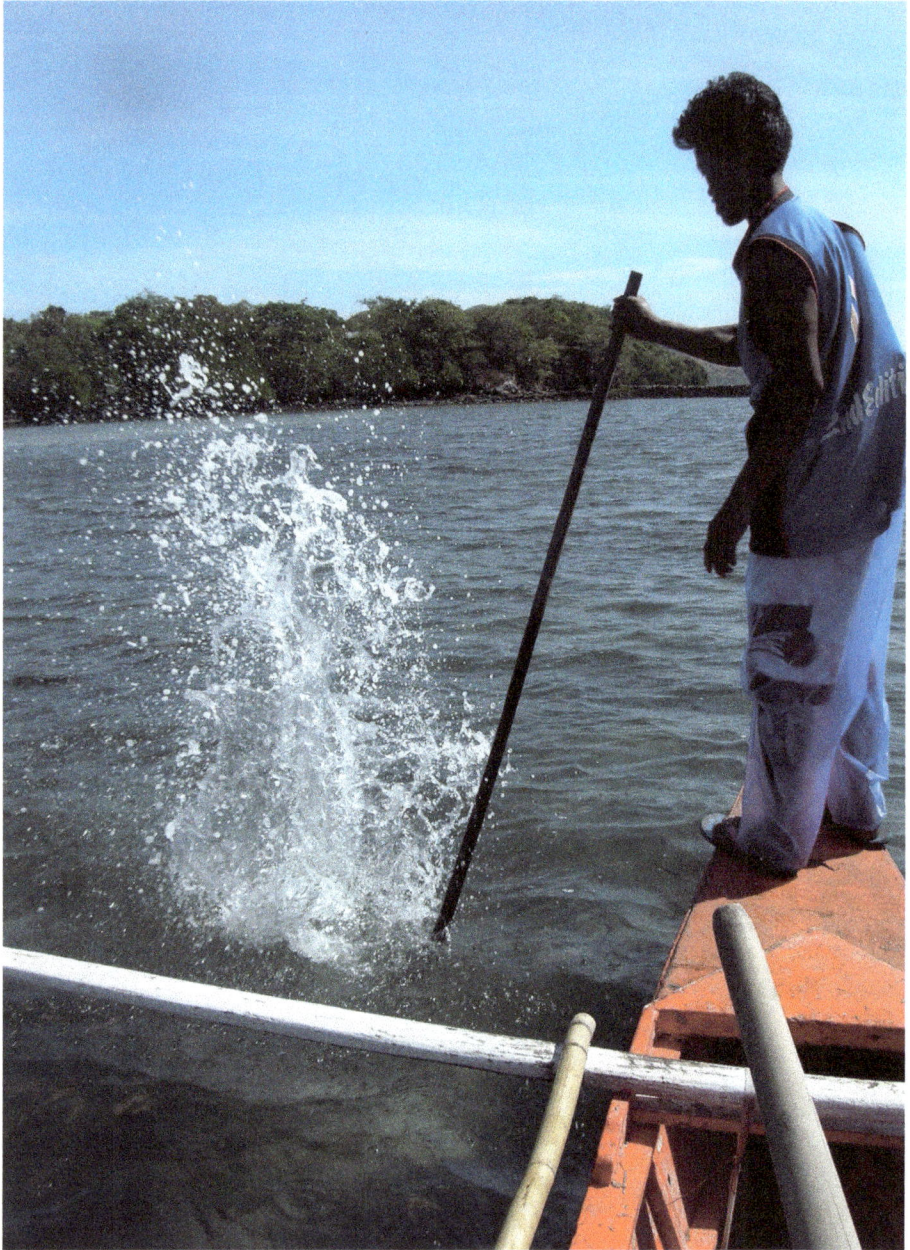

Plate 3-1: A fisher using the plunger to drive the fish into the gillnet.

Plate 3-2: Gillnet with the trip's catch.

Plate 3-3: Preparing net-caught fish to be dried.

Frequently, the owner of the boat is also the captain, but this is not always the case. Capital required for net fishing includes a boat, an engine and a net. In 2006–07 these costs would amount to between ₱40 000–50 000, depending mostly on the power of the engine and whether it was new or second-hand. Given the average monthly earnings for many net fishers is under ₱10 000, obtaining the money for this equipment is a significant task. As with other fishers who seek to buy a boat, most simply borrow money to do this either from wealthier kin and other personal connections, or through the Landbank[5] in Coron town. These sorts of loans are usually contractually based (Hollnsteiner 1970), and depending on the source of the loan, they are slowly paid back over time. Ed, for example, originally borrowed money from a variety of sources to pay for his net fishing boat when he first came from Bohol to live in Esperanza in 1989. Some money came from his brother Gary, who had already established himself in Esperanza several years earlier. Gary introduced two other creditors in Coron town to him; these creditors had originally helped Gary with his own net fishing boat. Ed paid monthly installments on the loan for several years until it was completely repaid.

A minority of fishers avoid such relationships in order to obtain capital. Vicente, for example, slowly saved up for his pump boat over a period of several years:

> I was extremely thrifty for a long time to get my boat. Sometimes, I didn't have any *ulam* with my dinner; I just left it for my family to have. I saved up little by little, and first I got the boat made, then I bought the engine and eventually I got the net. You see, I really didn't want to go into debt. I don't care if I don't end up like them [points to the house of his successful neighbours, the Marquez family], I just want to keep my family secure and do it all by myself.

An average net fishing trip of three to four days costs between ₱800–1200, depending mostly on the fishing location. Expenses include boat fuel, ice for fish storage and rice, coffee and cigarettes for the crew. The owner finances this trip, whether he is on the boat or not. The relatively low expenses mean that the owner is usually able to do this trip without needing to go into debt.

When the time came for marketing, every single net fisher I interviewed in Esperanza had only one buyer to whom they sold their total catch. For example, Ramon, a net fisher in Esperanza, had an agreement with Mily, a buyer based in Coron town. During the dry season of amihan, when the fish catch was very low and the demand was higher (forcing prices up), he brought the fish to Mily who gave him a good price. During habagat, however, the market was glutted with all sorts of fish and consequently the demand was very low. During this

---

5   Landbank was the only bank in Coron town during 2006–07. It is a government bank that has a priority in rural development and credit extension to the fishery and farming sectors.

period, Mily would only take the fish she could manage to sell; only then would Ramon go to other buyers to sell his fish. Thus, Mily had a regular supply of fish during lean times, and Ramon was guaranteed a steady price and a buyer of fish during periods of over supply. The problem, however, during habagat when there was little demand, was that Ramon had to wait until all of his fish were sold before he was paid. This had led him to become frustrated with net fishing during habagat, and he eventually turned to fusilier fishing during this time of year instead. So the economically personalised relationship between Ramon and Mily didn't necessarily always overcome all of the economic and environmental constraints.

The relationship between net fishers and buyers in Coron is marked by respect. Despite the fact that Ramon was several years older than Mily, he called her 'Ate Mily' (older sister Mily), a term of respect.[6] At the same time however, their relationship was not characterised by the same degree of distinction that characterises those between fishers and buyers in both the fresh and the live grouper fisheries. Mily and other buyers based in Coron town would often come down to the beach in Esperanza as the net fishers arrived with their catch in the mornings, laughing and joking with the fishers as they unpacked the fish from the nets. The sums of money exchanged are not as great as in other fisheries, and the subsequent class distinctions are not as strong.

The way in which profits are shared is quite simple, and is reminiscent of other accounts of share systems in Southeast Asia (Firth 1966; Stacey 2007)—one third goes to each of the two crew members, and one third goes to 'the boat'. Eder (2008: 70–1) reported that in San Vicente, if the owner of the net is not the owner of the boat, the net owner will also get a share of the profits. He points out that 'the principle used in determining compensation is that capital comes before labour. In the Philippines, capital commands a considerable share of the total revenue' (ibid.: 71). When I lived in Esperanza, there were no such ownership arrangements where the net owner was different to the boat owner. However, even when the boat and net owner are the same, as in Esperanza, the share system gives rise to a practice whereby the owner of a boat will often stay on shore, picking up a third of the profits while he hires his boat out to others to do the actual fishing for him. It is widely perceived that owning a boat, even one of the small pump boats used in net fishing, is a means to a reasonably steady income. Hence, being a boat owner entails relative social status and economic security.

For others, however, being forced to 'just join as a crew member' (*magsama lang*) has little security. As net fisher Don put it: 'Being a boat owner would be much better; that is my goal one day. Then I could relax at home sometimes

---

6   The use of fictive kinship is common in the Philippines; I expand on this theme in Chapter Seven.

while others went out and did the fishing for me! But now I have to fish for every single centavo I earn'. These fishers often move between different net fishing boats in the village, dependent on the schedules and requirements of the owners. Usually however, the fishers who do not own boats establish working relationships with two or three boat owners who are kinsmen, friends or neighbours. In this situation, fishers who don't own a boat become personally 'indebted' to the boat owners. They will attempt to repay this indebtedness by working hard and reliably on the fishing boat, and attending to any other tasks the boat owner may require. Don, for example, had a working relationship with two pump boat owners for whom he fished. He had to constantly ask these two boat owners what their schedules were and what days he would be able to work. Because he had been given regular work by these two owners, he said that he had an obligation to fish well for them, and ensure that they got a good return on their investment in him.

Sometimes younger fishers who are not yet established, or older, poorer fishers who are not as proficient as others struggle to get a place as a crew member on any of the fisheries. Nick, for example, had worked on several fusilier boats (during habagat) and net fishing boats (during amihan) for many years. Unmarried, he was not regarded as a particularly able fisher. Yet he was still able to find regular work on these boats, as there is an unstated form of obligation among boat owners to give work to those who need it and ask for it. This is not always easily given, of course, and this obligation to give work away is always in tension with the business and practical neccessities of fishing. Ricardo, for example, normally worked as a crew on his father's net fishing boat. He asked Carlos Marquez one afternoon if it would be possible to crew on his next fresh grouper fishing trip in Tawi-Tawi; adding he wanted the chance to prove his skills with the method used in fresh grouper fishing (cascasan). Carlos brusquely responded that the crew had already been decided and that it was too late to change anything. The examples of these crew members negotiating with different boat owners is simply used to demonstrate the different ways in which relations between crew, captains and owners play out.

Net fishing is a relatively low risk, low return fishery; the start-up costs and ongoing expenses associated with fishing trips are not as high compared to other fisheries. While personalised relationships between buyers and sellers are always present in Esperanza, it is frequently but not always necessary to go into debt to start fishing in this fishery. While practiced by some residents all year round, net fishing is most common during amihan; the off-season for fusilier fishing.

Fusilier Fishing

The primary fishing season for fusiliers begins in late-April, when the captains and boat crews will work together preparing the boats for the upcoming trips. Some boats need significant repairs or the replacement of old outriggers; others merely need a fresh coat of paint or a basic clean. There were five active commercial fusilier boats in the 2006 season in Esperanza; two boats had recently moved to the fresh grouper fishery, two had gone into cargo shipping and two were grounded because of a lack of funds. Repairs and preparations were done over a month period, timed for the boat to be ready for its first trip when the wet season rains begin.

Before departure, the crew gathers at the home of the boat in Esperanza. Twenty to thirty crew members sleep on the boat during the trips and fish the outer reefs of the Calamianes and the northern mainland. Four of the five fusilier boats fished the reefs of Dibangan during 2006, while one smaller boat fished closer to Coron. *Caesio cuning* (red-bellied or yellowtail fusilier) is the primary target species (Plate 2-7), but a variety of other fusiliers from the Family Caesionidae are also caught. They are caught in schools using hook-and-line baited with shrimp; although a variety of tactics are used by successful captains to obtain a good catch. This technique was adopted in a neighbouring barangay and became widely used in Esperanza from the late-1980s onwards. The fish are stored in a hold below the deck packed with ice and by preserving the fish already caught the fishers are able to fish for more than a week at a time.

The typical boat crew for fusilier fishing comprises of a captain and crew. Specialised roles exist within the crew such as the engine-operator, the ice-hold packer and cook (*kusinero*), but the majority of the crew are simply fishers. Trips usually last a minimum of ten days, and can extend for up to fourteen days if sufficient catch has not been obtained. Wives and children remain at home during this period, involved in household work and watching for signs of rough weather.

There is a great energy among the community on arrival of any of the commercial fusilier boats. People congregate in and around the boat, joking and laughing as the fish are taken from the ice-hold on the boat, weighed and repacked into freshly iced boxes on the shore (Plate 3-4). A tricycle driver, or sometimes a truck, is hired to take the boxes of fish to the port terminal in town. There, they are weighed and processed by Bureau of Fisheries and Aquatic Resources (BFAR) staff, before being transferred to any one of the cargo or passenger vessels that ply the waters between Coron and Manila. Upon arrival in Manila, they are unloaded by a paid caretaker, who then arranges for the fish to be sold at the wholesale fish market in Manila. From there, the fish are finally transported to the individual markets from where they are bought by consumers. Meanwhile, at some point after the unpacking process has been completed back in Esperanza, drinks are bought (usually by the family of the captain or boat owner), and the crew sit around, luxuriating in their return home and telling stories of their trip. This process repeats itself numerous times, with boats averaging between seven to a maximum of ten trips per year; with the overall aim of two trips per month between May and September. The number of trips for each boat depends on variable factors such as the weather, the availability of ice and the finances of individual boat owners.

Plate 3-4: Packing the fusiliers in boxes with ice.

The equipment required for commercial fusilier fishing is more expensive than for net fishing, mostly because the boats are simply much bigger (Plate 3-5). The first fusilier boats in Esperanza were adapted for use from boats originally built to use for the basnig fishery using bag or lift-nets. While they range in size and power, all the vessels are larger and have a far more powerful engine than the net fishing boats. Start-up costs for the boat and engine depend considerably on the exact size of the boat, when it was built and the strength and age of the engine, but costs normally are in the hundreds of thousands of pesos, as opposed to tens of thousands of pesos for net fishing or ordinary pump boats. Arrangements for the financing of this capital equipment vary. Carlos Marquez built his boat in 1980–81 for the purposes of basnig fishing, and this was done with the financial assistance of a wealthy businessman in town who owned many basnig boats.

Another owner, Geronimo, obtained a large loan from the local bank, Landbank. Most others combine savings with credit or loans obtained from various sources such as Landbank, kin, and patrons in town.

Plate 3-5: Fusilier fishing boat.

Because the trips are of longer duration than those of the net fishers, there are a far greater number of crew and the expenses for trips are correspondingly higher. A typical trip in 2006–07, lasting between ten and fourteen days (depending on the catch and the weather), required ₱70–80 000. In recent years, the prices of fuel and ice (the two most significant expenses) have been steadily increasing, while the price of the fusiliers has stayed around the same (₱40–70 per kg when sold in Manila at the wholesale fish market). Maintenance costs are also high. Combined with a decreasing catch in the primary fishing grounds, this meant that many boat owners struggled to fund their fishing trips in 2006. Two of the eleven boats that had been used for fusilier fishing in the past remained grounded throughout the 2006 season, unable to fish because of a lack of funds; a further two had their boats contracted out for cargo transport instead. Other owners who could not meet the funds required for fishing trips on their own had access to creditors (usually close kin or wealthier patrons based in Coron town) who could provide funds. Geronimo, for example, had five different creditors for each trip he made. Each creditor contributed ₱10–20 000, and if he had a successful trip he was able to pay all of them back immediately. On other occasions, the Marquez family would finance them.

Many fusilier boat owners spoke with distress about their current situation. During the late-1980s and for most of the 1990s, the fusilier fishery was essentially booming: trips were shorter than they were in 2006, fewer crew were required and more generally the prices for all expenses were down. Of the fusilier boats that were still operational in 2006, most of the owners either held narrow profit margins or were continually mired in debt. Most owners expressed a desire to transfer to either one of the export fisheries that their neighbours were clearly having so much success in managing. Geronimo, for example, said that:

> Fishing for fusiliers is not really good anymore. Back when I started in the 1980s, it was great, nobody had any debts and we only fished for short periods. But live reef fishing and fresh grouper fishing, they are the two methods that I want to get into now. I am organising this year to start live reef fishing because I know a trader, but for *cascasan* I need the personal connections. I need to get to know a financier who will pay for it.

Geronimo's experience illustrates the importance of personal knowledge of a financier in order to make these sorts of moves. Financing someone like Geronimo for either a fusilier-fishing trip or a switch to a new fishery is seen as a 'big help' (*malaking tulong*). Fishers like Geronimo thus often capitalise on the notion that richer people are expected to redistribute their wealth in ways that benefit other members of the community. Since his fusilier fishing had become less profitable in the last few years and he was unable to fund his own fishing operations, he asked the Marquez family help him to do so. Even though this was not the most profitable operation, according to both him and the Marquez family they continued to do so in part because of a sense of obligation (which Geronimo actively pressed) to help their struggling neighbour and the crew he employed.

In the five operational fusilier fisheries enterprises in Esperanza in the 2006 season, the captain of the boat was either the owner or a close kin of the owner. Crew was usually selected from within Esperanza and the neighbouring barangays or sitios. Many fishers use the kinship metaphor when describing relations at sea; the captain is the 'father' and the crew are all 'siblings'. It does not take long to make close friends in such circumstances, and the bonds formed by the often intense experiences at sea among the commercial fishers are particularly tight. What I observed in Esperanza is similar to what Russell (1997: 85–6) notes with regard to the Batangas fishing community she studied, where the shared experience of working at sea tends to form strong relationships of trust between captains and crew. Captains and boat owners (especially of the larger commercial vessels) of Esperanza typically belong to households of higher status and are generally better off, but still maintain common interests with their crews. The captains of the five operational fusilier boats were highly regarded within Esperanza.

The patterns of sharing profits in the fusilier fishery are slightly more complicated than how profit is divided among the net fishers. The captain automatically receives ₱4 for every kilogram of fish caught by the boat each trip. Other crews with specialised roles such as the engine-operator and ice-hold packers receive ₱2 for every kilogram of fish caught by the boat while the cook takes ₱1 per kilogram caught. In addition, they also receive payments for any fish they catch themselves. Each of the fisher crew receive ₱16 for every kilogram of fusiliers they catch, or ₱40 for any 'first class' fish they catch. This is a steady rate; independent of market fluctuations and the owner of the boat takes all the risks if there is a poor price for fusiliers in Manila. For example, if a boat catches 2000 kg of fusiliers, the captain of the boat will earn ₱8000, the engine-operator ₱4000, and each crew member ₱16 for every kilogram of fish he has individually caught (which, given a crew of 20, would equate to approximately 100 kg for each fisher and thus a payment of ₱1600). One person has the role of recording and weighing every fish caught by each fisher, so that the amount to be paid to each fisher is easily tallied at the end of the trip. Fishers also receive an advance payment (*bale*) of ₱500–700 before each trip; often used to buy things like cigarettes, clothes and other personal supplies for the trip.

The fusilier fishery was a fishery in decline by 2006; economic and environmental conditions no longer favoured it as well as they had in the 1990s. No longer able to fish outside of the wet seaon, many captains had turned (or expressed the desire to turn) to other gears and fisheries. All of the captains had gone into debt to establish their fishing operations, and many now needed to go into debt to fund each trip. Despite this, the fusilier fishery was still a prominent commercial fishery that provided employment for many people in the sitio.

Fresh Grouper Fishery

Unlike the seasonal fusilier fishery, the fresh grouper fishery operates all year round. The primary species targeted in this fishery is *Plectropomus leopardus*, which is exported to Taiwan, but there are a range of lower-value groupers and other reef fish that are caught as well. In 2006–07 some of these grouper boats operated in the far south waters of the Sulu Archipelago in Mindanao, near the Tawi-Tawi Islands. The crew was based at the provincial capital Puerto Princesa in between fishing trips. Three or four trips were conducted before the crew returned home to Mindoro and the Calamianes for a break. The technique of fishing for grouper using hook-and-line technology using a lure as opposed to bait (*cascasan*), was adopted in Esperanza by a captain who learned it from his relatives in Occidental Mindoro. With this technique, fishermen leave the mother boat on small one-person outriggers and fish by themselves in the open sea. They return to the boat at intervals to weigh and record their catch and then have it packed in the ice-hold. All of the fish are processed in the provincial capital Puerto Princesa before being transported to Manila, where a caretaker again processes them. The export value groupers are sold directly to an exporter based in Manila who subsequently sells the fish to an importer in Taiwan; the lower-value catch is sold at the wholesale fish market at Manila.

This fishery has only been in use in Esperanza since 2004, but the success of those who have adopted it has made many of those in the fusilier fishery eager to take it up. This success has been mostly due to the rising price of *P. leopardus*, fuelled by a strong demand in China. While this species is especially prized among Chinese consumers when freshly killed, it is always considered a delicacy even if it is killed at sea. In 2006–07 the price of fresh export-quality grouper was ₱500–600 per kg when sold in Manila. In all cases in Esperanza the captain was the owner, but the large capital necessary to run a fishing trip required the support of a financier.

There were two large boats involved in the fresh grouper fishery in Esperanza in 2006, and a fusilier boat was adapted for this fishery later in 2007. One boat belonged to the Calvino family and the Marquez family had the other boat as well as adapting a second boat for their grouper fishery business during 2007. While two of the boats were originally fusilier-fishing boats that had been adapted for use in the fresh grouper fishery, the first in the Marquez family fleet was purpose built for this fishery in 2004. It was considerably larger than the rest of the fusilier boats with two 90 hp engines accommodating up to 25 one-person outriggers (Plate 3-6). The Marquez family established contact with a Taiwanese fish export company (Sun Corporation) through their in Manila office, introduced by relatives based in Mindoro Occidental already involved in this fishery. Sun Corporation financed the ₱1 million required to build the boat, and some of the Marquez family's relatives from Mindoro joined as crew. The Marquez family was able to pay off this debt within two years. Carlos Marquez introduced the Calvino family to the manager of Sun Corporation (Ma'am Ping) during this period, and she subsequently financed the building of their boat as well.

Plate 3-6: Vessel used to catch fresh grouper.

The cost of each grouper-fishing trip is much greater than in the fusilier fishery for several reasons. Because of the local prestige associated with this fishery and the higher number of fishers, the total advance payments to the fishers are higher. In addition, because of the massive distance to be covered before the fishing grounds are reached, considerably more fuel is needed. Finally, the larger boat size has correspondingly higher maintenance costs. Around ₱200 000 was required to complete each trip, and this was financed by Sun Corporation for all three boats operating in 2007. The credit was subsequently paid off after each trip. Ma'am Ping was clearly regarded with a great deal of respect; an indication of this is the way Carlos, the head of the Marquez family dealt with her. One of the most respected and successful captains in Esperanza, Carlos was used to being the focus of respect from others. One of the few times I ever heard him use the Tagalog grammatical particle used when addressing someone with respect (*po*) was in conversation with her. Ma'am Ping did not visit the Calamianes, but stayed in Manila and all communication was conducted over the phone. Bill, the son of Carlos and Anna Marquez who stayed in Manila, would arrange transport of the fish from the Manila port to the Sun Corporation aquariums when shipments would arrive from Palawan.

The system of sharing returns of the fresh grouper fishery is essentially the same as for the fusilier fishery, except that the payments are much higher because the fish are much more valuable. Each crew member receives ₱110 per kg of export-quality grouper caught individually, and a further ₱16 per kg for any by-catch caught individually. As with the fusilier fishery, the captain and men performing specialised roles receive higher payments based on the overall catch of the boat. On an average trip, only one third of the catch may comprise export-quality grouper, while about two-thirds is a mix of by-catch. The captain would aim for around 300 kg of export-quality grouper; any less than 200 kg and the trip would make a loss. The captains and fishers are regarded with particular respect because of the distance and extra challenge involved in this fishery.

The two families involved in the fresh grouper fishery both stated that it had been a real blessing to them. Melinda Calvino, the mother of the Calvino household, told me that during their time in the fusilier fisheries from the 1990s through to 2005, the family steadily accumulated more and more debts. In just one year since transferring to the fresh grouper fishery, they had been able to pay off all of their debts. The only debt they had in 2006–07 was to Sun Corporation when they financed their trips, but they were able to make regular repayments after each trip; as a result, it was not understood as 'real' debt. At the end of 2006 they were able to enrol one of their sons in a mechanical course at a good college in Manila and enrol their daughter at the local university in Coron town. Melinda said that 'ever since we transferred to *cascasan*, we have been really blessed'. Similarly, the Marquez family told me how lucky they had been since transferring to the fresh grouper fishery from the fusilier fishery. Shortly after 2000, they were struggling financially, in debt and only able to run one of their two fusilier boats. Since transferring to the fresh grouper fishery, they had been able to pay off these debts and repair their other fusilier boat to work in the fresh grouper fishery as well. As the Marquezes described, there were simply many more fish in the Sulu Sea where they fished for fresh grouper, and the price for the groupers was much better than they were getting for fusiliers. These two factors made the fishery extremely rewarding financially. One particular fishing trip in the later part of 2006 had the whole of Esperanza gossiping furiously about the Marquez family hitting the 'jackpot'. Three tons of fish were caught—one ton of grouper suitable for export and two tons of mixed by-catch—yielding a profit of around ₱300 000; a story repeated enviously by the rest of the community.

The transfer to the fresh grouper fishery had not been advantageous in every way however. In particular, it required the crew to be based in Puerto Princesa for most of the year. Carlos' wife Anna often worried about his extended absences, and felt it was necessary to go there to visit sometimes. She felt that Carlos was getting old (he was over 50) and should have been thinking of an easier life based in Esperanza, managing instead of captaining the fishing trips. Economic necessities however, as she said ruefully, compelled him to keep fishing in the Sulu Sea.

The fresh grouper fishery is characterised by personal relationships between financiers and fishers. As Melinda Calvino said, '[e]veryone else would like to do *cascasan* as well but they need to be introduced to Sun as well. You need to have a financier to start doing *cascasan* because it is so expensive'. Again, then, the importance of being able to initiate a relationship with a financier is seen as vitally important to this fishery. Funding an operation like this is seen as helping a household, and households like the Marquezes and the Calvinos emphasised the importance of their personal relationship with Ma'am Ping in the success of their fishing. While other households had approached the Marquezes for an introduction to Sun, raising the value of obligation, by the beginning of 2007 the Calvinos remained the only family who had successfully done so. Not everyone could get an introduction of course, and the likelihood of economic success was of course a central factor in the Marquezes deciding which fishing families they would introduce to Sun and Ma'am Ping. Personal obligations were thus in tension with more practical factors.

Live Reef Fishing

Fishing for live reef fish is conducted on a modified version of the net-fishing boats described earlier, with an aquarium built into the hull (Plate 3-7, see also Padilla et al. 2003: 56–63 for other images). Once the fish are transferred to aquariums owned by a buyer in Coron town, they are flown to Manila and then on to Hong Kong, from where many are re-exported on to mainland China (see Plate 3-8). Although live fishing has been prevalent in the Calamianes for well over a decade, it only really began to increase in Esperanza from the late-1990s. Because of the extremely high price of live groupers (frequently above ₱2000 per kg return for individual fishers), this fishery continues to attract many fishers. Fishing is conducted all year round, but reaches a peak during wet season when the weather is more conducive to fishing. Aquaculture is also practiced (Plate 2-3). This is not sustainable full-cycle aquaculture, however; instead, fish are caught as sub-adults and held in cages until they reach a marketable size (0.5-1.0 kg). The global live reef fish trade is a highly significant trade (Scales et al. 2006), and dominates livelihoods for many throughout coastal Palawan (Eder 2008; Dressler and Fabinyi 2011; Fabinyi and Dalabajan 2011).

Plate 3-7: Boat used to catch live grouper.

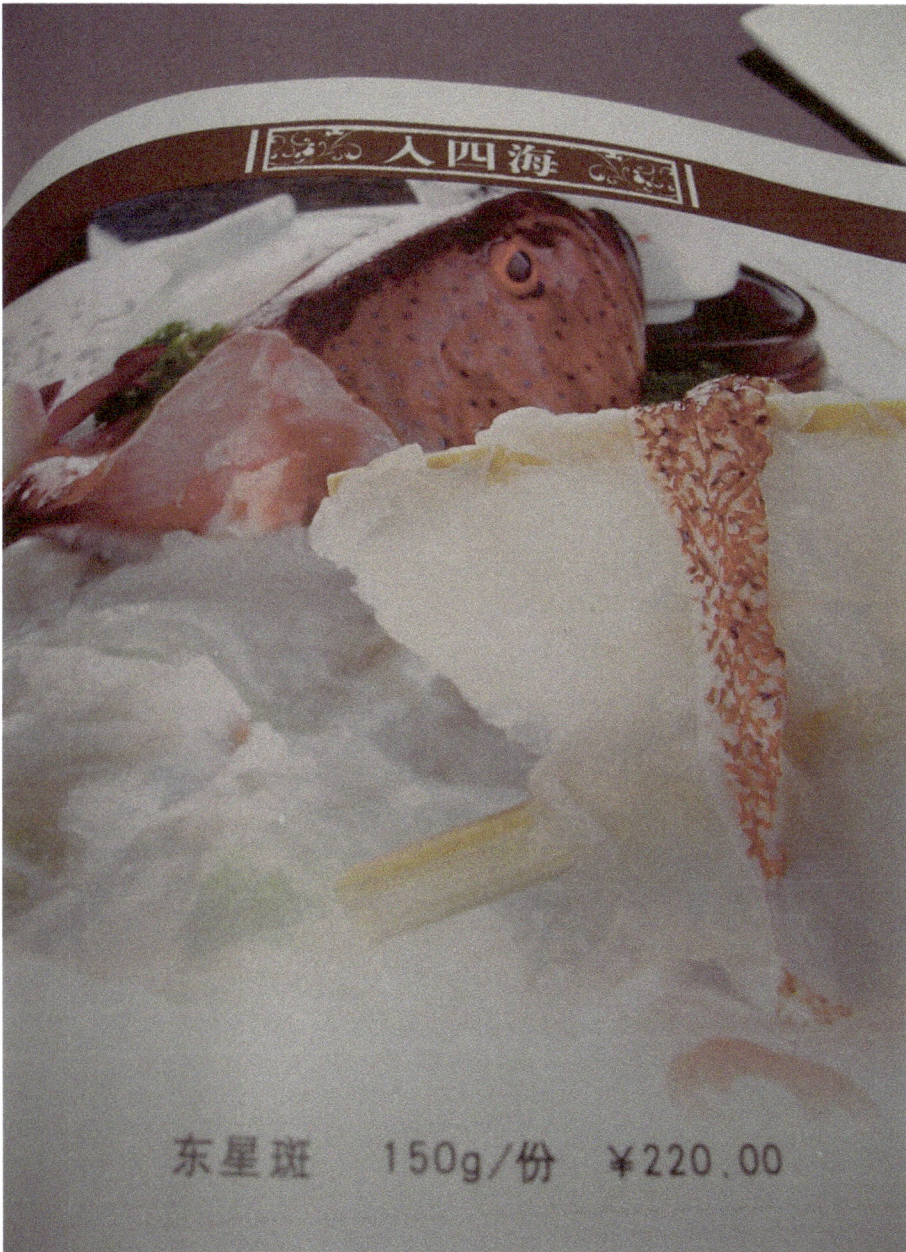

东星斑　　150g/份　￥220.00

Plate 3-8: Final destination of live grouper caught in the Calamianes waters: a Beijing restaurant.

The costs for the capital equipment for the live grouper fishery are somewhat more than those in net fishing as the boat is more specialised, and requires a 16 hp engine at the very least. A typical live grouper business set-up cost would range from ₱80–120 000 in 2006–07. Trips usually last longer than net fishing trips, and range much further. Sometimes a trip can take around four days if the catch is found quickly; more often, the trips last around ten days. Due to a sharp decline in grouper stocks in Coron Bay, most fishing is now done in more remote areas of the Calamianes or near the Palawan mainland (Padilla et al. 2003: 4; Fabinyi 2010). Because of the higher fuel and supply costs compared to net fishing, ₱5000–6000 was required for a standard trip during the 2006–07 period. All of the live leopard coral grouper caught in Calamianes waters is exported to Hong Kong (via Manila). When I interviewed eight restaurateurs specialising in live fish in Cebu and Manila, not one of them sold live leopard coral grouper anymore (some had previously), and it was only possible to buy the lower-valued species.

The live grouper fishery appeals mostly to fishers that were previously small-scale, typically using a net or simple hook-and-line from a canoe. While it is not regarded as a 'commercial' fishery under the regulations of the Philippines because the boats are under three gross tons, more money is involved both in initial capitalisation and in fishing trip maintenance compared to net fishing and hence it is heavily associated with credit extension. All of the live reef fishers I interviewed got their start in the fishery with the help of a financier, who arranged for the purchase and construction of the engine and the boat. Usually, a meeting between the financier (one of the buyers based in Coron town) and the fisher is set up through a mutual acquaintance. Often, this is another fisher already working for that buyer or sometimes someone better connected, such as a barangay councillor. Enrique, for example, owned his own net fishing boat and sometimes worked as a crew member on a fusilier-fishing boat during the wet season until 2004. At this time he approached a friend who had worked for several years for one of the live fish traders, asking him to act as a 'go-between' and set up a meeting with the trader. Enrique asked his friend to recommend him as a hard-working and successful fisher. As with the fresh grouper fishery, these are therefore not simple relationships between traders and fishers, but personal relationships facilitated by mutual acquaintances.

Once the relationship has formed, fishers like Enrique enter into a similar sort of personalised exchange relationship as described earlier with regard to net fishers. The major difference between these two patterns of financing however is that the fisher is usually deeply in debt to the fish buyer: firstly, with regard to the purchase and construction of the equipment, and secondly, with regard to the trip expenses. In practice, fishers ask their traders for personal loans and try to defer payments on the original loans as well. Credit is therefore often extended to a fisher and his family for his personal needs. In this way, the relationships resemble the 'patron-client' relationships more commonly seen in agricultural regions of the Philippines. The fisher is then required to sell all of his fish to his creditor until he has paid off his debt for the start-up costs. Once this has been achieved, the fisher is free to switch buyers or to enter into debt again with the first buyer for another boat. One of the traders described what he would do when fishers refused to play by these rules and sold to another trader; firstly, he would give a warning, and if it happened again then he would take the engine and boat back but he said in his experience this occurred infrequently. Crew of these boats originated from more diverse locations than the net fishery, and included people from neighbouring barangays.

Interestingly however, the biggest trader in Coron town during 2006–07 was the only trader who did not practice the credit-based system of personalised exchange. He avoided this mode of buying for several reasons. Firstly, before 2001 only some of the fishers in the Calamianes had the equipment needed to catch live fish, so many new players required considerable start-up capital, and needed credit for the boats and engines. But by 2001 when he started doing business in Coron, many fishers already had access to these things, so credit for the equipment was not always necessary. He said that often when a fisherman had debt, there was no pressure to pay off the original equipment debt 'because even after two to three years of debt they are still bringing their fish to you'. Eventually, he said, they would just 'forget' the debt because of the length of time lapsed— the practice of constantly bringing the fish to one buyer built a personal relationship to the extent that fishers felt they could get away with defaulting on the equipment loan. Even if the debt was paid off after a couple of years, he said that this meant that they were no longer bound to that particular buyer. The equipment debt in itself was not enough to bind a fisher to a buyer in the long term so after all the investment the buyer could still lose the fisher's patronage. To avoid losing the relationship developed with the fishers and instead of maintaining personalised economic relationships, he offered a better price than the other traders to fishermen (he called this the 'true price'). Additionally, he practiced *pakikisama* with the fishers by extending them small personal loans and allowing them to eat, sleep and drink at his aquarium on occasions.

For other fishers involved in the live reef fishery, relations of debt and credit have a large influence on their everyday lives. For some fishers, having debt to a financier is like carrying a huge burden. According to these fishers, having debt means that if your catch was small, you only have enough to meet your basic needs like food: 'Any extra profit that you make has to go to your financier to pay off your debt' said one live fish collector. Some are suspicious that buyers pay lower prices to fishers indebted to them because the fishers have no choice to whom they sell: 'If you cannot sell anywhere else, you have no choice. Then they give you the wrong price; there is nothing you can do' another live fish collector alleged. For some men, being in debt is viewed as a significant blow to independence and self-confidence. It can present in some as psychologically damaging, threatening their idealised role of what it is to be masculine. Successful fishermen are seen as men who have no problems looking after the family. The accounts of fishers like these and traders such as the one mentioned in the previous paragraph are an indication that personalised relationships are not necessarily harmonious and free of tension.

An important point to note is that in 2006–07 many of the live fish traders held extremely high status in Coron municipality. Two of the 'Big Three' live fish traders were also municipal councillors, and all of the others were either closely related to other councillors or at the very least, had good political connections. The lifestyles and personal influence of these traders were thus extremely different to even those of the elite of Esperanza. The wide social gap between traders and fishers was often marked by considerable deference on the part of the fisher. The last trader I discussed was noteworthy for the unusually comradely relationships he held with the fishers. In this context of marked social difference, the credit-based economic relationship tends to reinforce such wide social gaps.[7]

Organisations and individuals involved in attempts to regulate the live fish trade in Palawan have commented on the crediting and financing systems of the fishery, arguing that they encourage dependence and favour the traders over the fishers. A report commissioned by WWF to investigate the sustainability of the live fish trade in Palawan emphasised the dependency and inequitable distribution of benefits in the financing arrangements: '[w]ith returns uncertain, costs [of fishers] are compounded and the debts increase. To ensure needed support, they offer loyalty to operators, and a vicious cycle of dependence is perpetuated. Thus, fisherfolk become increasingly disempowered and marginalized' (Padilla et al. 2003: 92).

Another report on the live reef fish trade in the Calamianes by CI emphasised the problem of debt and its consequences. Referring to those fishermen working for financiers, it contended that:

> dependency is reinforced as debts accumulates. The dependence ensures the supply of live fish to the dealer, but leaves fishers powerless to act contrary to the dealers (sic) wishes and sell to other traders who offer higher prices. Indebted fisher folk practicing illegal fishing methods associated with the LRFT (live reef fish trade) are unlikely to

---

7   See Pinches 1991 for a discussion of how hiya is invoked in class relations.

be in a position to respect MPA establishment and supporting fisheries management rules if the dealers whom they owe money (sic) are not in favour of them (CI 2003: 9).

Finally, a policy brief by Pomeroy and colleagues acknowledged the profits from the live reef fish trade but stresses their short-term nature:

> The LRFFT (live reef food fish trade) has provided some communities with the opportunity to earn—at least temporarily—additional income from their fish resources in the area where very few income-generating opportunities exist. For the most part, these benefits have been gained in the short run with considerable long-run costs, ecologically, economically and socially. While many fishers have gained an income in the short term, in many cases they end up indebted to brokers, or required to fish in a way that is incompatible with local practices and habits (Pomeroy et al. 2005: 23).

Based on economic data collected during 2003–04, this policy brief also argued that these trading arrangements result in a significantly inequitable distribution of benefits. Based on figures of 2100 live fish collectors and 15 traders in the Calamianes in 2004, it estimated that each individual collector received 0.03 per cent of the total gross revenue of ₱475 905 500 of the Calamianes (ibid.: 86). Individual traders however received 0.87 per cent of the total— almost 30 times as much.

Many of those involved in the live reef fishery do not seem to mind the process of getting into debt and acknowledge that it seems to be a necessary element of succeeding in this fishery for most poor fishers. The crediting arrangements vary widely; some fishers had paid off the debt on one boat and then gone into debt again for a second boat that they could then hire out to other fishers; others had paid their debt on the initial equipment expenses but still occasionally went to their buyer for trip expenses and personal credit. Importantly, despite the associations with heavy debt, many of the live reef fishers expressed great satisfaction at the results of their livelihood. The majority of these fishers I interviewed had been net fishers or simple hook-and-line fishers without a pump boat before they had started live reef fishing, and so the greater rewards of live reef fishing had made a massive difference in their lives. Marco for example, simply pointed to his TV and stereo as purchases he had made in the last five years since he had turned to live reef fishing; purchases that he would never have been able to make before this.[8] All of them wished to continue with live reef fishing, viewing it as the most profitable fishery for fishers of their level of income.

---

8 Such a view recalls the observation by economist Joan Robinson that 'the misery of being exploited by capitalists is nothing compared to the misery of not being exploited at all' (Robinson 2006 [1962]: 45).

This optimism tempered during the latter part of 2006, however, when the issue of heavily regulating the live reef fish trade came to the fore (described and analysed in Chapter 7).

The views of the live fish traders based in Coron town invariably reflected such positive perceptions of the fishers mentioned above. All of them emphasised that the fishers' debt did not have to be paid-off straight away; that it was usually only paid-off in installments when the fisher had a good catch. As one live fish trader once told me as he raised a glass of rum to toast his fisherman counterpart sitting next to him, with a broad grin on his face: 'We are all winners in this fishery'. Indeed, the life histories of some of the live fish traders themselves reflects both the transformative power of this fishery in the Calamianes, and the potential for social mobility in the Philippines.

Joey was from a poor fishing family that had originally migrated from Bicol in the 1960s. Going to school in Coron in the 1970s, he had 'only one pair of pants' that he had to wear every day because all of his family's spare money was spent on his school fees. Starting with a small-scale dried fish business, Joey was one of the very first fish traders to get involved with the live fish trade in the early-1990s. Again, he was able to do this because of his wife's personal connections to an importer based in Hong Kong. Throughout the 1990s his business grew, so by 2006 he had around 500 boats fishing for him, based in the port area of Coron town. By 2006, he was one of the most influential politicians and successful traders in the municipality. Similarly, Bob was originally a fisherman himself from a rural part of Coron municipality who by 2006 had 300 boats fishing for him and had also made a successful second career in politics. Not surprisingly, he saw the live fish trade as 'a big help' to the people of the Calamianes.

Conclusion

I have aimed in this chapter to provide a window into how some of the relations between different groups in Esperanza function. As a simple description, my focus therefore has not been to provide a comprehensive discussion of the circulation of wealth, nor of all the strategies for economic and social security that households in Esperanza adopt. Instead, I have tried to show how relations between different groups in Esperanza tend to be mediated by patterns of economic personalisation. All these relations are bound up in an ongoing series of appeals, claims, and negotiations that form part of the 'everyday politics' (Kerkvliet 1990) that occurs in every barangay of the Philippines.

Poorer people in Esperanza relate to those better off or with more resources in a variety of ways. Frequently however, poor people try to set up personalised, reciprocal relationships, as I have shown particularly through the different relationships and exchanges involved in fishing. All transactions tend to require personalised connections between people of unequal status. Such connections include: providing capital to finance individual fishing trips, selling fish at a consistent rate, working as a crew member on a boat, switching to a new fishery, or in other contexts such as asking for help with matters like education and medical costs. The importance of managing good personal relations with those of higher status is seen as a fundamentally important way of going about everyday life. Having 'good' personal relations with someone of higher status does not necessarily always mean 'smooth interpersonal relations' (Lynch 1970), of course; I have tried to show how such relationships can encompass a range of pressures and tensions (such as the pressure to redistribute wealth).

These patterns of socio-economic relations I have identified in the fisheries of Esperanza thus foreshadow several themes highlighted and developed in the later part of this book. While the diversity of these relationships makes it difficult to generalise, frequently they involve certain cultural idioms that emphasise a sense of fairness for poor people. The next four chapters shift focus to examine how versions of these relationships are enacted in the discourse of the poor moral fisher. Here, I show how fishers take the underlying ideas about fairness for the poor, and reproduce them in a particular form during debates about environmental regulation.

4. The 'Poor Moral Fisher': Local Conceptions of Environmental Degradation, Fishing and Poverty in Esperanza

Kawawa kami ang mga maliit [Small people like us are pitiful] (Armando, a live grouper fisherman).

My style of fishing has no impact on the environment at all because it is legal. It is only the illegal types of fishing that damage the environment for everyone else. I use old techniques—hook-and-line—this is only enough just to survive (Geronimo, a commercial fusilier fisherman).

This second quote from Geronimo exemplifies two central features of local understandings about fishing and environmental degradation in Esperanza that were widely expressed. Firstly, so-called 'illegal fishing' is understood to be the prime factor behind environmental degradation; 'legal fishing' does not damage the environment or impact upon fish stocks. Secondly, these legal techniques only produce enough 'just to survive'; legal fishing is closely tied to poverty. Based on these two features, it follows therefore that any regulation of fishing practices should concentrate on regulating those who are doing the damage to the environment, and those who can afford to be regulated.

This chapter will explore in detail such local ideas that fishers express about fishing. It will fuse two ethnographic themes: one is their representation of the causes of environmental degradation; the other is their representation of their own fishing practices as moral, humble and tied to poverty. These two ideas are brought together in what I call the discourse of the poor moral fisher. Linking analyses of local representations of the environment, morality, fishing and poverty, I will argue that this discourse constitutes a particular understanding of the relationship between poverty and the environment. In Esperanza, poverty, a lack of impact on the environment, and morality are seen as mutually reinforcing; conversely, wealth, environmental degradation and immorality are understood as related in the same way (as illustrated in Figure 4-1).

LEGAL FISHING   **ILLEGAL FISHING**

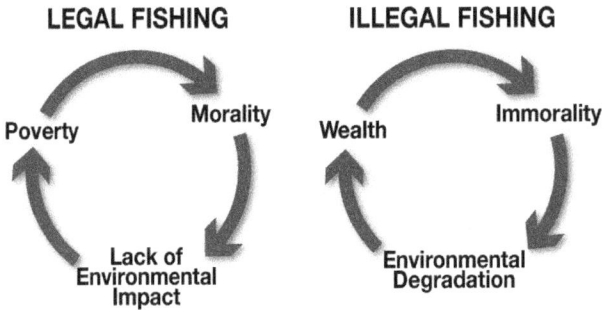

Figure 4-1: The discourse of the poor moral fisher.

Linking Environmental Degradation, Fishing, Morality and Poverty

Politicised Environmental Knowledge

After situating my analysis in a discussion of various sets of literature, the second section of the chapter details the understanding of many fishers that the state of their marine environment is degraded. I examine the causes fishers ascribe to this degradation, emphasising their strong distinctions between legal and illegal modes of fishing. I demonstrate how such attributions are closely connected to deeply-held views about morality and legality in the Philippines.

This part of the chapter draws upon the work of writers such as Filer (2009) and Brosius (2006), who emphasise how we should consider local environmental knowledge as a form of political knowledge. Arguing with regard to the Penan of Malaysia, Brosius contends that '[w]hat matters is not how much Penan know about the landscape they inhabit, but how they position that knowledge, and themselves, within the broader contours of power' (ibid.: 136). In Esperanza, local knowledge about the marine environment is more complex than simply environmental knowledge. It cannot be divorced from the political and moral claims in which it is embedded. Here, local knowledge of ecology and the marine environment is closely intertwined with particular social claims.

Elsewhere in Palawan, Eder (2008) has described the existence of similar attitudes towards the attribution of blame for the problems of environmental degradation in San Vicente. There, Eder reports (ibid.: 112), a strong distinction is made by locals between highly efficient and active operations, such as beach seining, and 'truly' illegal activities such as cyanide and dynamite fishing. Local residents argue that the truly illegal activities are the ones that the authorities should concentrate on regulating, as they are the ones that do the greatest amount of damage. Eder also found that among the fishers he interviewed, 'lingering problems with illegal fishing practices in San Vicente could be blamed primarily on intruders from other locales rather than on local residents' (ibid.: 118). He goes on to argue that:

> [a]t a discursive level ... this frequently uttered proposition may be deployed to another truth and absolve local residents of responsibility for environmental damage. If outsiders are the cause of the damage, then they are the ones who should be monitored and reminded of the environmental rules. Local people tend not to see themselves as the ones who need monitoring (ibid.: 119–20).

Similarly, as I will show in the Calamianes, local knowledge about fishing is also 'political' knowledge, strategically deployed by residents in an effort to strengthen their claims. However, here I extend Eder's description of this phenomenon to investigate the cultural frameworks that underlie this reasoning. A focus of my argument is that linked to these claims is the representation of legal fishing practices as moral, humble and tied to poverty.

Understanding Poverty and Morality in the Philippines

As illegal fishing is represented as a particularly immoral activity, legal fishing is seen as moral, in part because of its allegedly negligible effect on the environment, but also because of its close association with poverty. Hence the discussion of this chapter will focus on the close associations between legal fishing and poverty, arguing that fishing is widely viewed as an arduous, difficult occupation with very low social status. While in Chapter 3 I detailed some of the more objective characteristics of poverty, livelihoods and class relations, here I focus on subjective representations about fishing and poverty.

The latter part of this chapter builds on the work of Cannell (1999), who has written about a related, widespread and powerful discourse in the Philippines that is focused on pity and sympathy for the weak and poor. Cannell argues that poor people in Bicol present themselves as pitiful in order to enter into reciprocal relationships—the sorts of reciprocal relationships I identified in Chapter Three. She identifies a number of writers who have previously analysed these attitudes in different locations of the Philippines (ibid.: 231–4). Polo argued that in Leyte, fishermen used the idiom of pity to obtain help in cases of misfortune or hardship. A fisher seeking a loan would say: 'How pitiful I am today. I will have to borrow something from you' (Polo 1985: 56). In her ethnography of markets in Panay, Blanc-Szanton (1972: 129) argued that fishers claimed that '[e]veryone has a right to survive and provide for his family—a right which transcends all other economic or legal considerations'. In this case, fish dealers and other fishers were obliged to share with those who had been less successful, despite the lack of any clear economic incentive, 'based on what is regarded as a fundamental human right to some minimal income with which to feed oneself and family, and the obligation of one's fellows to provide it if possible' (ibid.: 123).

Kerkvliet (1990) found the existence of a similar ethic among poor peasants in Central Luzon, who assert that 'they are as human beings entitled to security and dignity. And if necessary, those needs and rights should be satisfied at the expense of others who have more than enough to assure their own fundamental rights' (ibid: 269). While Kerkvliet suggests that this 'basic rights discourse' may have originated partly from the upheavals of the peasant Huk Rebellion in Central Luzon of the 1950s, he also emphasises particular features of Tagalog language and culture that may have contributed as well (ibid.: 272–3). He notes that classic accounts of Filipino culture (for example Hollnsteiner 1963; Lynch and De Guzman II 1970) have focused a lot of attention on more passive values such as *pakikisama* (the ability to get along with people) and *hiya* (shame or shyness), while ignoring various terms in Tagalog that convey far more active values related to the discourse outlined above. Kerkvliet (1990: 273) lists terms such as: rights (*karapatan*); equality (*pantay-pantay*); dignity (*karaŋgalan* and *pagkatao*); humanity (*makatao*); justice (*katarungan*); freedom (*kalayaan*), the unity of self with others (*kapwa*); and treating others as equals, as human beings (*pakikipagkapwa*).

Drawing on the historical work of Scott (1983) and Rafael (1988), Cannell (1999: 237–9) suggests that these conceptions may have their origin in pre-colonial forms of social relations between commoners and the aristocrats. Since Cannell's work has appeared, the work of Junker (1999) has provided a highly detailed analysis of pre-Hispanic political economy in the Philippines that lends support to this view. Junker emphasises that the power of local leaders was obtained through command of labour (as opposed to land), and that this meant that leaders had to attract and maintain followers through developing personal ties of reciprocity. As part of these dyadic relationships, leaders were obliged to fulfill reasonable requests from their followers for assistance (ibid.: 238).

Cannell (1999: 233) also makes the important point that '[it] would be as impossible as it is unnecessary to disengage many Filipino lowland terms from the history of Christianity'. It is perhaps enough to point out the powerful role of the Catholic Church in the Philippines and the Calamianes, with its associated morals of noblesse oblige and pity for the poor. Indeed, the 'Church of the poor' that the Catholic Church of the Philippines supposedly represents is explicit about this identification. Borchgrevink (2003) has similarly drawn attention to the ways in which the notion of sacrifice is a central idea throughout much of the Philippines: 'Briefly put, the idea is that by undergoing some form of hardship, one will be able to help somebody else. The underlying idea is one of exchange:

by doing something for God (or for a saint)—praising, honouring or suffering for him somehow—it is expected that he will do a service in return' (ibid.: 55). From this perspective, hardships and suffering because of poverty are related to sacrifice, and lend an air of nobility to poverty (see also Wiegele 2006: 502).

Former President Joseph Estrada is one public figure who was able to successfully appeal to this discourse focusing on the poor, and whose story illustrates the potency of this discourse in the Philippines (Plate 4-1). Using the nickname 'Erap',[1] Estrada's career is littered with references to pro-poor rhetoric: from his 1998 campaign slogan of *Erap para sa mahirap* (Erap for the poor), his portrayal of movie characters who identify themselves with the underdog, to his declaration upon winning the Presidency: 'The common people have waited long enough for their turn, for their day to come. That day is here' (cited in Doronila 2001: 2).

Plate 4-1: Fernando Poe Jr (Presidental candidate in 2004) and Joseph Estrada (President 1998–2001) forged film careers playing characters identifying with the poor. *Tatak ng Tundo* was about life in the slums of Manila.

---

1   *Erap* is *pare* backwards; a shortened form of *'compadre'* and used by Joseph Estrada to identify with the common man. In colloquial Filipino *pare* means mate, buddy or pal.

Similarly, Rafael (2000) has analysed how Philippine national hero Jose Rizal's

> potency rested on his ability to evoke populist visions of utopic communities held together by an ethos of mutual caring, the sharing of obligations (*damayan*), and the exchange of pity (*awa*).... It was precisely this image of Rizal in conjunction with the suffering Christ—figures at once pathetic and prophetic—that was mobilized to explain the events that began with the assassination of Ninoy Aquino in 1983 and ended with the People Power Revolt in 1986 that ousted the Marcoses from power.... Predicated on the logics of suffering and sacrifice, the political culture of People Power and subsequent regime of Cory Aquino thrived on the notion of pity rather than equal rights to legitimate their claims to power and moral certainty' (ibid.: 211–2; see also Rafael 2010: 173–4).

From 2005–10, television viewers could see the discourse of 'pity' played out every day on the game show *Wowowee*, which included a segment that involved the host questioning contestants about the circumstances of their life. Contestants, frequently coming from poor backgrounds, would routinely break down in tears as they described their difficulties. Moved by their distress and the accompanying emotive music, audiences would take pity on the contestants, and (depending on his mood) the host would offer them cash bonuses.

In the context of environmental regulation, the emphasis that NGOs frequently place on social justice in the Philippines (Austin 2003; Bryant 2005) is indicative of this perspective. Indeed, Bryant (2005) argues that a central strategic goal of NGOs in the Philippines is to accrue what he terms 'moral capital'. While Bryant does not focus on the meanings of morality in the Philippines, I would suggest that the quest for moral capital among these NGOs is particularly pertinent in the Philippines because of the themes I have just surveyed.

The discourse of the poor moral fisher, I shall show, can be situated within these broader discourses of pity and sympathy for the poor in the Philippines. Fishers represent their practices as environmentally harmless, as difficult, humble, and closely tied to family and poverty: they are therefore deserving of pity. This next ethnographic section takes up this theme by detailing how illegal fishing is represented as the sole cause of environmental degradation.

Illegal Fishing as the Cause of Environmental Degradation

Perceptions of Environmental Degradation

As I described in Chapters 2 and 3, fishers in Esperanza have employed an extremely wide range of techniques and targeted many different species over a period of several decades. Lower fish catch has obviously been a key factor in the shift of fisheries activities in response to the various fisheries booms in the Calamianes. The shift has been characterised by basnig boats targeting scads and anchovies in the 1970s and early-1980s, to the largo boats targeting various first class fish in the 1980s, to the fusilier fishery that boomed during the 1990s, and lastly, the live and fresh grouper fisheries. This history, combined with the increasing level of inputs required such as more crew, more fuel for longer and further trips, and smaller mesh sizes for net fishing suggests that most fishers would have an understanding that fish stocks have changed over time.

When asked about the state of fisheries at the time of my fieldwork, most fishers had very similar views. In a survey I conducted in 2006 among 60 fishers of various types in Esperanza, 47–78 per cent—stated that there were fewer fish in their fishing grounds compared to when they first arrived to the sitio. The fusilier fishery that was so profitable in the 1990s was seen by many as a struggling fishery without a future 15 years on. One experienced fusilier-fishing captain told me that the fishery had 'maybe five or seven years left' in it before everyone else turned to either the live or fresh grouper fishery. Fishers complained that the number of crew required for a successful fishing trip had risen, that trips had to go for much longer and further away, and that expenses for fishing trips also rose while the price of the fusiliers had remained the same for a long time. Net fishers similarly complained of the increasing length of trips they had to finance just to break even: 'It used to be we could just go there (pointing to the shore of Esperanza); or to San Andres (the neighbouring sitio to Esperanza). Sometimes we have to go to Bulalacao, Busuanga. We have to sleep on our boats sometimes; we have to stay three days sometimes now'. By 2009, the live fish trade was also seen as a fishery in decline. The area around Coron was widely seen as over fished, and those fishers who had the equipment and capital to do so went further afield (mostly to Linapacan) to richer fishing grounds (Fabinyi 2010).

When talking of the past state of fisheries in the Calamianes, many fishers often commented nostalgically on the ease of fishing. 'Before, you could catch fish just inside here (pointing to Coron Bay)' one fusilier-fishing captain observed. 'You would only need to go fishing for three or four days, then you would have enough fish'. Now, he described, most commercial fishing captains had to fish for a minimum of ten days to get the required amount of fish, and fishing had to be conducted further and further away. Many fishers related how fishing nowadays was thoroughly enmeshed within debt cycles. For them, the lack of easy fish catch had made their lives much more difficult. Fusilier fishing captain Geronimo, for example, said that:

> When I first arrived here in the 1970s, life was good. I was fishing all year round (*basnig*), and even though we didn't have much money then, life was easier. Starting in the 1980s though, *basnig* became weaker, and then by 1990s only fusiliers were good to fish. Now though, even fusiliers are weak now. We can only fish during *habagat* for fusiliers, and even then we have to fish further away—you can't fish in Coron Bay anymore; you have to go to Dibangan, Liminancong, Dumaran (locations closer to or on the Palawan mainland). And I have to fish for much longer, at least ten days every trip now. Soon everyone will need to switch to the live fish trade or the fresh grouper fishery. And before, during *amihan* we used to fish *largo*, or even fusiliers still. But now I just do net fishing during this time.

Like Austin's (2003) research with fishers in Honda Bay, Central Palawan, I found it difficult to get reliable quantitative information about the extent of environmental degradation in the area. This was due to both the unreliability of government statistics and the imprecision of fishers' recollections. While the fishers' nostalgia for the good days of the past may be exaggerated, it is nevertheless a key component of the discourse of the poor moral fisher. My survey results and interview data make it clear that most fishers—of all gear types—had a strong perception that fish stocks were declining in their fishing grounds. Many of the net fishers suggested that over the last ten years in particular stocks had been declining especially fast. Common estimates of average catches of rabbitfish for a one-day trip, for example, ranged from 20–40 kg, compared to 50–70 kg ten years before.[2]

National and Local Variations of 'Illegal' Fishing

The notion of 'illegal' fishing is a contested one in the Philippines, and can reflect a range of activities that occur in Philippine waters. Perhaps most commonly, the term describes the activities of commercial scale vessels that intrude into municipal waters. Numerous studies have documented the prevalence of such activities in the Philippines (for example Russell and Alexander 2000). Such vessels have included the Danish seine or *hulbot-hulbot*, purse seining vessels

---

2 Eder reported similar declines among fishermen in San Vicente, where estimations of fish catch fell from an average of 30–50 kg a day in 1980 to about 10 kg a day in 1997 (Eder 2008: 101).

and various sorts of trawlers. In Palawan, this is often associated with the activities of foreign vessels arriving from the South China Sea. Several recent cases have been heavily publicised in the Philippine media. In December 2006, the Chinese MV Hoi Wan was caught in the UNESCO World Heritage Site of the Tubbataha Reefs in the Sulu Sea with more than 1500 live fish, including more than 350 CITES protected, endangered Napoleon Wrasse (Jimenez-David 2007). And in September 2007, another Chinese vessel was found to be holding more than 200 turtles (mostly endangered green sea turtles), 10 000 turtle eggs and two live pelagic thresher sharks when boarded for a routine inspection (Wildlife Extra 2007).

More recently in 2008, the Palawan fisheries authorities encountered numerous cases of Vietnamese fishers fishing around southern Palawan (Anda 2008a, 2008b). Commercial fishing in inshore areas of the Calamianes was common in the 1970s, and Wright (1978) declared then that more serious than the problem of dynamite fishing was 'the indiscriminant [sic] inshore trawling practiced by vessels from Manila which destroys plant and animal life in the shallows, and the physical environment that supports life' (ibid.: 123; see also Fabinyi forthcoming). During my stay in 2005–07, less commercial fishing was conducted in areas close to shore, except for municipally registered vessels. At night, however, the glow of the baby purse seiners were sometimes seen, especially in more remote areas of the Calamianes where they would enter into the 15 km municipal zone. Since 2008, a large commercial operation, called *lintig* (a smaller version of *muro-ami* fishing) had been operating in inshore waters around Coron with the cooperation of the municipal authorities, an issue that had stoked resentment among some local residents, especially those who fished inshore waters.

Dynamite fishing is another common form of illegal fishing in the Philippines, and Wright (1978: 123) reported its existence in the Calamianes back in 1978. The technique involves using a glass bottle filled with dynamite; the lit wick is thrown into the water near schooling fish, causing an explosion that kills or stuns the fish, making it easy to scoop them up. Dynamite not only kills all of the nearby fish life but also causes great damage to coral reefs. In more remote parts of the Calamianes, where the level of law enforcement is low, dynamite is often alleged to be used in various types of fresh fisheries. During 2006–07, several exporters of fresh fish to Manila in the Calamianes were found with fish killed by dynamite on board their vessels.[3] In August 2007, three sacks of ammonium nitrate, an agent used in explosives, were found on a cargo ship bound for Coron while in port at Manila (GMANews.TV 2007).

---

3   Unlike fish killed using cyanide, which requires a chemical test to determine the presence of cyanide, fish killed with explosives are often easier to detect as their internal flesh and organs are frequently damaged and bloody. (Fishers often disguise this by salting and drying the fish.)

Cyanide fishing is most closely associated with both forms of the live fish trade in the Philippines: the live reef fish for food trade, and the marine ornamental aquarium trade. The technique is quite simple: cyanide tablets are dissolved in water, which is then squirted on to reef, stunning the fish, which are then easily collected. Cyanide also destroys corals (Cervino et al. 2003). Dalabajan (2005), an environmental law expert in the Philippines, estimates that there were around 250 000 cyanide fishing trips between 1999 and 2002 in the Calamianes. He argues that the reason for this large number is the incapacity of the local governments to enforce the relevant laws.

Accusations of corruption in fisheries governance were routine among fishers I interviewed on the issue. According to Padilla et al. (2003: 87), local fishers 'indicated that the reason why illegal activities persist was the strong link between unscrupulous traders/operators and law enforcement agents'. Furthermore, some conservationists were concerned that two of the three biggest live fish traders in Coron town were also council members of the municipal government.

Finally, there are a range of fishing practices which are often technically illegal but are usually far more openly contested. These include the use of certain active gears that may be locally regulated, such as beach seining or the use of fine mesh nets. Fishing within protected areas, while technically illegal, is not viewed with the same degree of abhorrence by local fishers, and as I show in Chapters 5 and 6, it is a highly contested issue. Usually, when people talk about 'illegal fishing' methods in the Calamianes they are referring to one of two types of fishing: the use of cyanide or, less frequently, dynamite.

Illegal Fishing is to Blame

In the survey I undertook of the fishers in Esperanza, 36 of the 47 fishers who stated that there was less fish catch in their fishing grounds since arriving in Esperanza—or 77 per cent of the 47 who gave this answer—went on to say that the primary cause of this situation was because of the activities of the illegal fishers (*illegalistas*).[4] When returning from their commercial fishing trips

---

4   Agrawal asserts that transformations in patterns of resource use among rural communities in Kumaon, India, are dependent on two related beliefs: '(1) nature is an entity discrete from humans and endangered by reckless human actions; and (2) this endangered nature needs protection, which can be generated in the form of careful government' (Agrawal 2005: 201). Once these beliefs arise, he argues, '[t]hose who see the environment as requiring protection are more likely to put greater effort in their protectionist practices' (ibid.: 163). Fishers in Esperanza certainly have a very real sense that environmental degradation is occurring as a result of human actions, and, as I detail in Chapter 7, fishers also feel very strongly about the need for 'careful government' and wise regulation. However, fishers represent the causes of environmental degradation as only the result of 'some' human actions and not others. While fishers in Esperanza may hold both of the beliefs very strongly that the marine environment can be damaged by human actions, and that the marine environment requires protection, the beneficial patterns of resource use that Agrawal suggests will follow on from these beliefs have not occurred. This is because of the ways that fishers allocate responsibility for environmental degradation, asserting that illegal fishers only are to blame.

to Dibangan and Tawi-Tawi, fishers would relate how they frequently see other boats out at sea, fishing with dynamite. While fishing captains in the fusilier fishery would be happy if they caught around two tons of fish during a regular trip, they said that it was easily possible to catch up to three or four tons of fish using dynamite. They emphatically declared, however, that they did not use dynamite themselves, because it was illegal. Similarly, fishers in the live reef fish trade would complain that all of the corals in their favourite fishing grounds were damaged because of the use of cyanide.

Carlos Marquez, for example, said that he was very 'proud' that he had never used illegal fishing methods:

> Often when I go fishing in Dibangan, I see boats from Batangas, from Lucena … they use dynamite to blow up the fish. I get very angry because it spoils the reefs for the rest of us. I tell them it is banned, but they don't listen…. To me, the difference is between short and long term thinking. Those who want to get rich quickly, who want to make lots of money and don't care about anything else, they will go for illegal fishing. But if you want to think about your children, and if you want to do the right thing, you will stick to legal fishing.

Similarly, fusilier-fishing captain Geronimo angrily declared that:

> it is the illegal fishers who have wrecked the good fishing grounds for everyone else. Out in Reed Bank (in the South China Sea, location of largo fishing) now the place is no good anymore…. All these Vietnamese, Thai, Chinese, Taiwanese boats have used trawlers and dynamite to damage them. And here too, even in Calamianes, cyanide has been used everywhere. Maybe it is the poor people with ten children who do that sort of fishing, but the problem is with the bosses, with the drug lords.

Many other fishers also resented the way that the government would let the 'drug lords' fish illegally with impunity. Fusilier fishing captain Manuel remarked sarcastically, for example, that illegal fishers from other countries were 'welcome in the Philippines'. The perception was that while legal fishers tried hard to do the right thing, and suffered in poverty as a result, illegal fishers became rich and went unpunished. Such resentment became more overt during the provincial government's attempt to impose regulations on the live fish trade, the topic of Chapter Seven.

While fishers in Esperanza squarely blamed the illegal fishers for environmental degradation, they denied any suggestion that their own activities may have been damaging to the environment. One captain, for example, asserted that 'if everybody used legal methods, the fish won't get finished'. When I asked Geronimo about whether he thought his fishing activities could potentially be harmful to the environment or reducing fish stocks, he confidently declared that this was not possible, responding with the quote I placed at the beginning of the chapter. As I describe in detail in Chapters 5 through to 7, these sorts of perspectives meant that in the opinion of the fishers, any regulations formed by the government should focus entirely on cracking down on illegal fishing. Informants would also sometimes acknowledge or concede that part of the problem had to be simply because of the large influx to Esperanza and the Calamianes of fishers of all types, illegal and legal, over the last 30 years. The extent to which illegal fishing was consistently to blame (both among survey respondents and through everyday conversations) for all reef degradation and stock depletion, however, was enough to strike me as particularly significant.

Fishers did not necessarily base these assertions about the relative damage of legal and illegal fishing on detailed or thorough understandings about ecological processes. Understanding of ecology among fishers was notably limited. Questions about seasonal migration patterns of fusiliers, for example, would often elicit vague, inaccurate and wildly varied responses. Some of these included fusiliers: migrating 'to another country' or warmer waters; swimming to deep water to feed on different prey during the non-fishing season; diving to the bottom of the sea to lay their eggs during certain periods; and to become hungry because of a lack of their normal food during the fishing season and thus be tempted by the bait of the fishers.[5]

The radical dichotomy between the damage caused by legal and illegal fishing that fishers expressed bears a resemblance in many ways to what Johnson (2006) has described as discourses that 'valorise' small-scale fisheries, similar to the idealisation of traditional ecological knowledge. Johnson argues that such discourses ascribe values such as social justice and ecological sustainability to small-scale fisheries (ibid.: 747). He demonstrates in his article that '[t]hese powerful but not always explicitly acknowledged valuations do not necessarily correspond to the reality of small-scale fisheries, which can be exploitative and ecologically destructive' (ibid.). Similarly, a recent report by the WWF has argued that 'to say simplistically that "small-scale" fisheries are "low impact" is a false and dangerous generalization' (WWF 2008), citing numerous examples from around the world where small-scale fishers have been the primary causes behind resource depletion and environmental degradation.

---

5   This is not to say that fishers have little local knowledge—on the contrary, fishers demonstrate an exceptionally deep knowledge of aspects of fishing including knowing where to fish, the use of successful techniques and understanding of many different weather conditions. Instead, what I observed in Esperanza is similar to what others writing about local marine ecological knowledge have detected (Foale 1998; Pollnac and Johnson 2005; Sabetian and Foale 2006). Foale (1998: 200), for example, has argued that '[m]uch of the local knowledge possessed by subsistence, artisanal and commercial fishers is focused on how to locate individuals of a target species in space and time, and, once located, how to capture them'. Such knowledge can be detailed and impressive, but frequently lacks information about specific ecological relationships such as reproductive biology and stock dispersal (Sabetian and Foale 2006: 8).

Data from NGOs about the extent of environmental degradation in the Calamianes has focused on the effects of the live fish trade. Reports from WWF, CI, and the FISH project all emphasise the high level of overfishing and stock decline in the grouper fishery. Reporting on the extent of damaged coral cover, Padilla et al. (2003: 52) found high percentages of dead coral in their surveys in 2001, ranging from 20–100 per cent in selected sites.

Importantly, they make the point that 'it is difficult to assume that all "dead corals" are in fact caused by cyanide' (ibid.: 50), noting that many other factors may have also contributed. They point out that '[d]etecting the effects of cyanide on coral reefs at the Calamianes has been difficult' (ibid.: 53) because of the possibility of other factors, and that overfishing more generally certainly has many causes. Their overall conclusions and recommendations for management stress the broad problem of overfishing as the primary causal factor. They attribute this problem of overfishing to many causes, including increased migration to the Calamianes, a lack of stock quotas or other effective regulatory measures such as protected areas in spawning locations, and the apparent prevalence of cyanide use. Similarly, a CI report in 1998 attributed the parlous state of the fisheries in the region to a combination of illegal and legal methods (Werner and Allen 2000: 60–2). The authors assert that the basic problem of overfishing, while partly caused by cyanide, is also fundamentally because of the sheer increase in fishers in the Calamianes over a long period.

In Esperanza, therefore, it is likely that assertions illegal fishing is solely to blame for all environmental degradation, while legal fishing is not, are not based on detailed ecological analyses about the impact of fishing on stocks. When questioned about how, exactly, cyanide and dynamite fishing had reduced stocks whereas other techniques such as hook-and-line had not, fishers would point to the increased capacity to fish using illegal methods, and the damage that they did to corals, which they often described as the 'home' of the fish. These two points are of course correct—illegal fishing clearly does contribute to overfishing and the destruction of corals. That legal fishing, however, is represented by fishers as completely harmless, in such contrast to illegal fishing, points to other forces and ideas at play. Instead of being simple accounts of ecological change, an important point to note about discussions of illegal and legal fishing in Esperanza is the ways in which they are framed within discourses of identity, locality and morality.

Identity and Legality

Illegal fishing is a 'hot' political issue in Coron, one with many murky layers of gossip among various actors. When asked who these illegal fishers actually were, fishers and other residents of Esperanza would always refer to another group of people. When it was fishing within the marine protected areas near Esperanza under discussion, always fishers from town or from another barangay were blamed. Similarly, cyanide fishing was only ever conducted by those outside the barangay, according to anyone within the barangay. At a larger geographical scale, when people talked about dynamite fishing in the more remote, outer reaches of the Calamianes and Palawan waters, it was always boats from Marinduque, Lucena and especially Batangas that were cited as the culprits. The illegal fishers more generally were depicted as unscrupulous, thirsting for easy money at the expense of the smaller fishermen. In the words of Raul, a net fisherman during amihan and crewman on commercial fusilier boats during habagat who had lived in Esperanza since the 1970s: 'These people, they come from Marinduque and Mindoro, and they fish everything here. These people all live in town; they will just do something else or move on once all the fish are finished. But we, the fishermen here in Esperanza, we have no option but to stay here and keep fishing'. Similarly, live reef fish collector Roly argued that 'these transient fishers, they come from the Visayas to here because this place Palawan has the most fish out of all of the Philippines. They come here because there are no fish left in the Visayas; they have finished them already'.

I argue that part of the explanation for this response lies in concepts of identity and morality. This argument draws upon the work of Walley (2004) in Mafia, Tanzania, and Zayas (1994) elsewhere in the Visayan Islands. In a similar case, recent migrants and other 'socially marginal individuals' (Walley 2004: 55) were blamed for dynamite fishing. Walley argues that the social networks and sanctions among the original inhabitants or owners discourage them from participating in dynamite fishing, whereas those without such networks 'are more likely to engage in activities that are personally profitable but are an anathema to Mafia residents' (ibid.: 56).

In the Visayan Islands, a similar insider/outsider dichotomy can be observed among coastal residents. Here, sojourning fishermen (*pangayaw*), are contrasted with local residents (*tumandok*) (Zayas 1994). Zayas has described how in many parts of the Visayas, sojourning fishermen will migrate to different islands for fishing, dependent on the seasons of amihan and habagat. These migratory fisher groups construct and maintain social relationships with local residents, exchanging fishing knowledge, technology and capital for the rights to fish in the area. She speculates that as resources are depleted, it would be more difficult to maintain these relationships. Versions of such relationships continued at the time of my own fieldwork in various parts of the Visayas, including Negros Oriental and Cebu (personal communication, Magne Knudsen and Shio Segi, May 2008).

In the Calamianes, the Cebuano terms *pangayaw* and *tumandok* are rarely used in everyday conversations, at least not by the Tagalog-speaking fishermen I spoke with regularly; but there is a common distinction between people who are 'from here' (*taga rito*) and 'migrants' (*mga dayuhan*) or 'not from here' (*hindi taga rito*). Which category someone belongs to depends completely on the speaker's perspective. Many local residents of Esperanza, for example, would confidently assert that they are 'from here', arguing that they had been there for more than 30 years, or among the younger population, that they had been born there. For many of the Cuyonon residents, the expatriate population and the indigenous Tagbanua population around Coron, all Visayan migrants (the vast majority of whom have migrated since the post-war period) are 'not from here'. At a broad level, it is these 'Visayans' who are most frequently blamed for illegal fishing by these residents of Coron.

Among the fishers I worked with however, there is a sub-category of 'Visayan' concerning fishers in the Calamianes who are transient, in the manner of the sojourning fishermen described earlier. Such fishers 'keep coming back' (*balik-balik*). A small island off Coron Island is host to one such community of transient Boholanos who first started to visit the area about 1986. At the time of my fieldwork they moved between the island off Coron, an island in neighbouring Culion municipality and their original home of Bohol. Fishers elsewhere in the Calamianes and environmentalists allege many of this community's fishers practice cyanide and dynamite fishing. Similarly, many of the residents of Esperanza were originally transient fishers, especially coming from the Waray-speaking region of Northern Samar. As reports filtered back to Samar in the 1970s about the relative success of residents who had already migrated to Esperanza, acquaintances and relatives from Samar came and fished during the habagat monsoon. Some were transient for many years before settling down permanently in Esperanza.

This pattern of intermittently fishing and then settling down to live permanently in the region has been common for a long time, and transient fishing continued during the period of my fieldwork. Wright (1978) reported that many of the migrants who eventually settled in the Calamianes originally appeared there as crewmen on fishing vessels from Manila. By 1978, Wright states, '[i]tinerant fishermen in numbers up to about 2000 are always present' (ibid.: 66). In Coron town and the surrounding barangays when I was living there, other groups of small-scale transient fishers would sometimes appear, especially during the prime fishing season of habagat between May and September. They would report to the barangay captain, who would usually give them permission to stay (for a fee), in a manner similar to that outlined by Zayas (1994) in the Visayas. Similarly, some baby purse seining vessels would obtain permission from individual barangays to operate in nearby waters for a season at a time.

This would also usually involve negotiation and a payment of some kind to the barangay, such as payment for a specific infrastructure project (such as, building a jetty or basketball court).

Many fishers, however, fish transiently for periods of time without necessarily basing themselves at a particular location. On relatively small-scale vessels, fishers will still be able to make trips of several days at a time, so they can come from various locations around Northern Palawan to fish in the Calamianes. Esperanza fishers, for example, will frequently fish for live grouper without formal municipal permits on reefs in other parts of the Calamianes, or in northern mainland Palawan. Unlike when fishers base themselves in a particular location, or when fishers conduct gillnet fishing in locations very close to shore, such (small-scale) fishing in open water appears to be regarded as relatively open-access by other fishers. As one fisher put it, talking of his experiences fishing for live grouper without a permit in Linapacan, 'they are not that strict for this type of fishing…. We are all fishers and we understand that you need to go to different locations to fish…. It is only when people start using illegal methods that other fishers start getting angry'.

Since 2005, however, the Coron municipal government has been attempting to reduce this sort of fishing through the introduction of registration for all municipal fishing vessels. This is in large part an initiative of the FISH project. Under the proposed legislation, all boats, gears and fishers will have to be registered with the municipality, and greater powers will be given to fish wardens from the barangays to enforce these laws. The primary aim of this new legislation is to make it more difficult for transient fishers to fish in Coron waters.

One possible line of argument is that transient fishers are probably responsible for a lot of the illegal fishing. Without the complex sets of social networks that long-term residents had, such fishers would potentially feel more able to conduct socially disruptive activities such as illegal fishing. As I will argue in Chapter 6, young men from the Calamianes who do not see their long-term future in local fishing, and who are under greater pressure to demonstrate their economic prowess, are also more likely to participate. This situation of relating illegal fishing tensions to tensions about social identity therefore bears a close resemblance to that described by Walley (2004).

A different argument, however, would be that transient fishers are not actually responsible, but that they are simply easier to blame.[6] From this perspective, it may be a case of the situation foreshadowed by Zayas (1994) in the Visayas; where the local residents simply decide not to allocate as many rights to incoming

---

6 See for example Knudsen (in Fabinyi et al. 2010: 623–5), for a case in Negros Oriental where length of residence is shown to be a powerful determining factor in social status and the apportioning of blame for illegal fishing.

migratory fishers as fishing pressure increases. Whether transient fishers are actually responsible or not for illegal fishing, however, is not really the point. What is interesting here is the ways in which outsiders are framed as responsible; these discourses about social identity thus infuse any discussion of illegal fishing with an explicitly strategic perspective. As with the situation described by Eder (2008: 119–20) in San Vicente, ascribing the causes of environmental degradation to the activities of outsiders means that the activities of local fishers should not be regulated.

This contrast and tension between recent and long-term residents mirrors in some ways the more commonly described tension between small-scale or artisanal fishers and commercial fishing boats. In the Philippines, such tensions have been commonly described, especially within the fisheries management literature (for example Russell and Alexander 2000; Green et al. 2003).[7] In the Calamianes, however, this tension is not as clear as in many other parts of the Philippines and the developing world. Instead of being a relatively straightforward case of contestation between small-scale and commercial fishers, fishers (both small-scale and commercial) in the Calamianes that I worked with are more concerned about the distinction between legal and illegal fishers. As I indicated earlier, by 2006 commercial fishing within municipal waters of the Calamianes was not generally considered as problematic as it used to be. The discourse of the poor moral fisher, therefore, is not only another version of the same complaints about large-scale commercial fishing that small-scale fishers have made across the developing world (see Bavinck 2005). In addition, this discourse is also concerned with more locally specific notions about what illegality actually means in the Philippines.

Morality and Legality

Depictions by residents in the Calamianes of illegal fishers as unprincipled and predatory outsiders resonate with how the practice is conceived in terms of morality. People frequently link the prevalence of illegal fishing to the prevalence of degenerate behaviour more broadly in local society. One local municipal councillor, for example, identified the central, indeed the only significant problem in Coron to be illegal activities: 'Coron is a very good municipality now. It will become a great municipality once it eliminates these illegal activities. Illegal fishing and drugs are the only big problems Coron faces'. Fishers liken illegal fishing to a vice, like gambling or drinking, which are 'easy' but immoral ways to spend your time. Because it is illegal however, this means that it attracts far worse condemnation than other vices.

---

7   Indeed throughout the developing world, this is one of the most commonly observed types of fisheries conflict. As Bavinck (2005: 806) points out, '[the] issue of conflict between small-scale and modern fisheries … continues to color the everyday experiences of millions of fishers in Asia, Africa, and Latin America'.

The act of illegality seems to transform the behaviour from an environmentally damaging fishing technique into a transgression against society. This is similar to the way many other illegal activities are represented in the Philippines. Sensationalist media coverage of drug problems tends to depict drug-takers simplistically, as moral deviants who have betrayed their family and society. There is no complexity or 'grey area' here; they are simply condemned as morally wrong. From a symbolic perspective then, the representation of certain types of fishing as 'illegal' takes on greater significance. The emphasis on the term 'illegal' can be seen partly as a trope or symbolic reference to underlying fears of other illegal activities (Geertz 1973: 193–233). Illegality is identified as extremely morally wrong; and because cyanide and dynamite fishing are illegal, they become morally wrong.

Such a moralistic perspective shapes the way that illegal fishers and fishing are analysed. This moralism has been linked by Mulder (2000: 184) to the way the public sphere is commonly represented in the media, literature and education in the Philippines: '[t]hrough the ahistorical and anti-sociological treatment of the twentieth century, history and society silt up in a muddle, and no intellectual tools are handed down to develop systematic comprehension of the wider world. Often moralism substitutes for reason. To save the country, we need a moral revolution!'. The fact that legal fishing can also be (and inevitably is) damaging to fish stocks is obscured by this moral perspective.

I have shown here how conceptions of illegal fishing as the sole cause of environmental degradation are not necessarily based on empirically demonstrated realities that show the effects of illegal fishing compared to legal fishing. The ideas of illegal fishing as morally wrong and environmentally destructive, and legal fishing as morally right and harmless, are not only based in ecological understandings, but also in tensions between new arrivals and long-term residents, and in broader conceptions about morality.

While illegal fishing is associated with immorality, wealth, and outsiders, legal fishing relates to morality, poverty and family. In order to develop this point it is necessary to appreciate how fishing in the Philippines is understood as an occupation closely linked with poverty. The following discussion looks further into the discourse of the poor moral fisher, by examining the ways in which fishing is represented by fishers and others in Philippine society as a livelihood marked by hardship, uncertainty and low status.

'Legal' Fishing and Poverty

The Difficult Nature of Fishing

Fishers in Esperanza represent fishing as a particularly difficult occupation, and one which is not a 'real' profession. Instead, fishing is often viewed and practiced as a frontier establishment strategy, or as a 'last-resort' livelihood. This section of my argument critically engages with the arguments of Pollnac et al. (2001b: 532), who assert that '[t]he assumption that fishers will readily shift to alternative occupations' is based on untested beliefs about fishing as an undesirable job, and fishers as the poorest of the poor. Based on survey results from the Philippines, Indonesia and Vietnam, they conclude that '[i]t is clear that in all three countries, fishers like their occupation and only a minority would change to another occupation, with similar income, if it were available' (ibid.: 541).

Pollnac et al. (2001b) are certainly correct to critique the occasionally naïve assumptions of fisheries managers who have sometimes taken for granted the ease of transferring fishers to alternative livelihoods. They also cite evidence that in many countries fishers are not the poorest of the poor. In many parts of the Calamianes, it is also true that inland farmers are understood to be the poorest people. However, I suggest in this section that Pollnac et al.'s (2001b) conclusions about the high levels of job satisfaction fishers experience, their positive views of fishing, and their unwillingness to change occupation may be stretching the point somewhat. Unfortunately, the quantitative surveys undertaken by Pollnac et al. (2001b) do not provide the in-depth qualitative data necessary to analyse the more complex ways fishers understand their practices. Another survey of fishers in Coron using similar simple questions, for example, found that a majority would be willing to change occupation (Baum and Maynard 1976: 40). While these two studies obviously addressed different times, economic conditions and specific locales, the point is that fishers will respond in a variety of ways to formal survey questions about alternative livelihoods, and that generalisations such as those of Pollnac et al. (2001b) do not necessarily always apply.

In my view, a more helpful approach is offered by Eder (2003), who, writing about coastal communities in mainland Palawan, emphasises the ways that:

> fishing itself, as a livelihood option, is always located on a wider field of other livelihood options. And just as not all fisherfolk used to fish for a living, not all remain fishers. The numerous former fisherfolk I encountered are a reminder that fishing is something that is also abandoned in favour of other occupations, both within the lifetimes

of individuals and—importantly—across generations in 'fishing' households … fishing was [frequently] not a career or a lifelong occupation so much as a household's establishment strategy, in much the same fashion as swiddening or *kaingin* may function as an establishment strategy on an agricultural frontier (ibid.: 215–6).

Eder also stresses that rarely is fishing the sole livelihood of coastal residents in countries like the Philippines.[8]

I have found similar practices and attitudes towards fishing among the residents of Esperanza. Often, fishing is not seen as a traditional 'occupation' like working in an office or as a labourer, but as simply a last-resort means to get by. When I conducted my first household survey at the beginning of my fieldwork, one of the questions I asked concerned what profession people held. Frequently, people would simply tell me that they had no profession. When I rephrased the question to ask how they got their income, they would then look at me, as if I was stupid, and tell me 'fishing, of course'. As I discussed earlier in Chapter 2, many of the residents who had moved to Esperanza did so not necessarily because they had been fishers in their earlier lives throughout the country. Bong, for example, had been a farmer in Panay for many years before turning to mining, and, when he moved out of that job, moved to Esperanza in the hope of succeeding in fishing, where he was at the time of fieldwork the engine-man on one of the commercial fusilier boats. He saw fishing as a central establishment strategy adopted relatively easily in a foreign province where he had few connections: 'If you don't know many people, then fishing is the easiest job to start with'.

Vicente, another fisherman, had been based in Manila for many years working as a labourer. When he found the expenses of this lifestyle too overwhelming, he came to Esperanza to live with his cousins and began net fishing: 'Manila was too difficult, my rent was too high, there are so many pressures there. I found myself hungry at times. So I decided to come and work with my family here. The only job I can do here is fishing though'. Similarly, Danny was a younger man who had originally fished on his father-in-law's fusilier boat. He moved to Manila for several years in the hope of making more money. After three years working in a factory with little money saved, he returned to Esperanza: 'I didn't really succeed in Manila, so I came back to work with my father-in-law on his fishing boat'. Young men of Esperanza, in particular, view fishing as something that is hopefully only going to be temporary (see Chapter 6). Commercial fusilier-fishing captain Manuel summed up the situation of many in Esperanza when he said that 'many people weren't fishermen before they came here. They came here just to look for a better life'.

---

8   Acheson (1981: 291) has cited various studies from around the world that concur with this point.

While fishers view their work with pride and clearly take a measure of satisfaction in it, as Pollnac et al. (2001b) rightfully point out, they also consistently emphasise the harder and more challenging aspects of the work. Fishers in the commercial fusilier, fresh and live grouper fisheries in particular, cite the lengthy trips (up to two weeks at a time) as challenging; mentally and physically. Bad weather, cramped and uncomfortable sleeping conditions, lack of access to fresh water for bathing, simple food and the emotional stress that comes with spending long times away from the family are just some of the hardships experienced by these fishers. When I first arrived in Esperanza, my requests to join these trips were met with comments like 'It is difficult! You will find it too hard'. Most small-scale fishers also emphasise the physical difficulties such as sitting in the open sun at sea for hours at a time, or of fighting the challenging weather and large waves. As I describe in more detail in Chapter 6, overcoming these difficulties with style and bravery is a sign of masculinity and status, but there is little doubt that fishers view their work as difficult and arduous as well.

Geronimo's description of his recent life history is a reminder of the difficult lives of many fishers in this community:

> I run up high debts with my fishing now. Each of my [fusilier fishing] trips costs about ₱80 000 to put on, and I have five different creditors who finance my trips. They each give me around ₱10–20 000. It is much harder to find financiers now; previously everyone was willing to help out but now maybe only five per cent of people are willing to give loans. Most of the profits I make now just go to my creditors; it's like I'm just working for them. Now I'm just getting by: I have enough money to eat and run my fishing trips, but not enough to keep my child in school in Panay and not enough to look after my [75 years] old mother in Panay. Now I'm very keen to move to the live fish trade, at the moment I'm looking for a financier to get me the engine. I want to send my child to college.

> But still, it could be worse. In 2002 my child drowned in a boat accident. Then one of my other children decided to stop studying, then my other child got sick with appendicitis. He was sick for weeks and I spent all my savings. Then I hurt my back and I didn't have enough money to see a doctor for myself. I couldn't even walk for one year and fifteen days; only now I am completely physically fit again. So since this bad time, I cannot recover all of my debts and life is much more difficult.

This is not to say that fishers never come to enjoy their occupation, find it fulfilling, or are always able to transfer to another job. Fishers would talk with pride about their skill as a captain and their knowledge and experience at sea.

Importantly, social status within the community is often determined by fishing skill and success. However, the point remains that while fishers may ascribe status, pride, and satisfaction to their livelihood of fishing, it remains a difficult and low-paid job.

When describing some of their hardships, fishers would often explicitly use the idiom of pity (*awa*). Fishers would describe themselves as pitiful (*kawawa*) during the difficult fishing season of amihan, for example, or with reference to a bad catch. Conversely, an old woman suggested once that God had 'taken pity on her' and given her a long and healthy life. With reference to illegal fishing, fishers would often talk of themselves as being pitiful because of the actions of illegal fishers. For example, holding up ten fingers, one older fisherman said that 'out of every ten people in Palawan, five are good and five are bad'. Closing the fingers of his right fist, he went on, 'these five are pitiful, but the other five—ay! They are rich!'

A key feature of fishing that make fishers view it as especially difficult is its uncertain nature. In an older review of the anthropology of fishing, Acheson (1981: 276) pointed out that '[f]ishing takes place in a very heterogeneous and uncertain environment. This uncertainty stems not only from the physical environment, but also from the social environment in which fishing takes place'. Various anthropologists in different countries have documented characteristics of the uncertain and risky environment that fishers must endure that include: the sea being an alien and dangerous environment; it is difficult to see fish (even with the aid of fish-finders); a lack of property rights; long hours; and the psychologically stressful nature of the work (ibid.: 276–7). Acheson (1981) also identifies various strategies fishers have adopted, such as share payment systems, an egalitarian emphasis, flexibility in crew recruitment, patterns of determining access to fishing rights, the development of personalised relationships with middlemen, and the use of ritual and magic.

An important element of local ideas about fishing in Esperanza (and elsewhere in the Philippines) is its association with the concept of 'luck' (*suwerte*). The role of luck and uncertainty does not necessarily distinguish the worldviews of fishers in the Philippines from those of fishers in many other countries, but it contributes to the sense that fishing is a difficult occupation. The uncertain nature of fishing, and the fisher's reliance on luck, is key in understanding how fishing is associated with difficulty. Szanton (1971: 60) for example, has argued that among fishers in neighbouring Panay during the 1960s, their worldview was shaped by a strong sense of fatalism: 'Good luck will yield riches, bad luck will deny them, and there is little one can do about it. Wealth will come, if at all, when it is in the nature of things to appear'.[9]

I do not interpret the comments about luck made by fishers in Esperanza to reflect the same degree of fatalism that Szanton saw as present among fishers in Panay during the 1960s. Captains of the commercial boats, in particular, tend to ascribe a greater level of relative success or otherwise to their skills, knowledge and experience, as well as to luck.[10] However, I do contend that most of the fishers in Esperanza, and certainly all of those associated with the smaller boats, have a very strong sense of the role of luck in determining their place in the world. Fishers frequently complain of the uncertain nature of their income—some days they return with little or no catch; and despite the skills and knowledge of experienced fishers, sometimes even they are unsuccessful. On many days during the windy season of amihan, fishers would wait glumly in their houses, waiting for the weather to change. While fishers attribute a significant degree of success to the captains' skill and knowledge, fishers are often simply unlucky. 'Jackpot-jackpot lang kami'[11] was how one live fish collector characterised his lifestyle, meaning that fishers simply move from jackpot-to-jackpot, with little security or certainty. Because of the highly variable nature of its income (see Chapter 6), hook-and-line fishers in particular feel this insecurity, yet I argue that in Esperanza all fishers feel a significant degree of uncertainty in their occupation.

This is apparent not only in comments fishers make about their lifestyle or the empirical range and fluctuations of fish catches, but also through the manner in which religion is expressed. Russell and Alexander (1996: 437–8) explain that in the Batangas fishing community they studied, amulets, magical rituals and baptisms from the Catholic Church are all employed by fishers as means of ensuring better fishing success. Polo (1985) has documented in detail how fishers in Binlayan, Leyte, constructed and practiced elaborate rituals invoking the spirits and people of the sea. Similarly, in Esperanza, fishers regularly call upon religion. Although I did not observe the use of amulets or magical rituals in fishing, boats are baptised, complete with godparents who act as advisors to the boat. Sermons usually involve pleas for protection by God, and prayers

9   Veloro (1994) has also illustrated, drawing from a case study in Southern Palawan, how *suwerte* in the Philippines can mean luck as in good fortune or chance, or it can also refer to an attribute of a person, as in with reference to their 'fate'. Small-scale fishers in one Palawan *barangay* Veloro was writing about attributed success in fishing to *suwerte*.

10   This, I would argue, is because of what is described in maritime anthropology as the 'skipper effect' (Pálsson and Durrenberger 1990; Russell and Alexander 1996). This effect refers to whether or not there is the perception that the role of a skipper plays a significant role in determining the size of the catch. Following Pálsson and Durrenberger (1990), and Russell and Alexander (1996) in the Philippines, I found that this sort of effect was present among more capitalised fleets with less personal labour recruitment—the commercial fusilier and fresh grouper boats.

11   This is a 'Taglish' (mix of Tagalog and English) phrase meaning that fishers simply live off 'jackpots'.

are often said before a trip. Often, fishers simply say that whether they have a good catch or not and whether their life is looked after or not, is simply left in the hands of God and reflects their level of luck. This uncertainty is seen as something that makes fishing more difficult than other occupations: 'Farming is something you can do every day, but fishing … what can we do when the weather is bad, when a storm comes? We have to be lucky if we are to do well in fishing', said Ray, an older commercial boat captain.

Plate 4-2: Fishing in the hot sun.

Plate 4-3: Fishing village shoreline.

The Low Status of Fishing

Many Filipinos of various occupations around the Calamianes represent their incomes as poor when compared to foreigners. Acutely aware of the large inequalities between the Philippines and developed countries, especially America, fishers frequently discuss international exchange rates. 'One US dollar, fifty pesos!' they would exclaim. The West in particular is perceived as a region where everyday hardships do not exist—people are able to pay for medicine to cure them when they get sick; if one can't find work, the government pays them, and people can eat whatever they want wherever they want. Foreigners are generally perceived as having a life of unattainable luxury and leisure. This perception has important implications when it comes to dealing with tourism.

Fishers in particular, however, usually depict their occupation as earning considerably less than other occupations in the Philippines, and are very humble about their level of social status. 'I'm just a fisherman' (*Mangingisda lang ako*) was a commonly heard refrain in the time I spent in Esperanza. Fusilier fishing captain Manny, for example, told me how he trained in radio technology during the 1970s but was unable to find a job, so he became 'just a fishing captain'.

This perception of fishing as an inferior occupation is also validated and reinforced by many other Filipinos. Many richer Filipinos view fishermen essentially as simple, poor peasants—especially Visayans, the linguistic group of most fishermen in the Calamianes. Even many richer Filipinos such as conservationists or government workers who work with fishers, and are more sympathetic towards them, still view them in a paternalistic manner. As Russell (1997) found in a coastal Tagalog community, while fishing captains may be respected figures and informal holders of power within the community, their continuing associations with manual labour, and their lack of wealth and family connections, mean that they are still regarded as having lower status in the wider society of the Philippines.[12]

In such a status-conscious country, the experience of being poor can often be degrading and humiliating. Often it is not simply the experience of living in material hardship cited as the worst thing, but the lack of dignity that comes along with this. Pinches (1991: 174) has argued, for example, that in the squatter settlement where he worked 'what matters most to people in Tatalon is the way others attribute or deny value to them as human beings'. Poverty is thus understood as 'the experience of not being valued as human beings' (ibid.: 177) in addition to the lack of material wealth.

---

12   In mainland Palawan, Austin (2003: 168–9) has also described a similar perception among small-scale fishermen, who feel powerless to change their lowly status in Philippine society.

When talking about their standard of living and understandings of poverty, fishers in Esperanza would usually strongly contrast the quality of life among Filipinos with that of foreigners. As foreign countries are perceived as rich decadent paradises, so the Philippines is seen as the most poverty-stricken place on Earth. 'There is no opportunity here in the Philippines. Even though my daughter has a college degree in computer science, she cannot get a job', one frustrated mother told me, in a version of an extremely common refrain. Getting a job overseas is highly valued and sought after among many younger residents, although out-migration among residents in Esperanza is not as common as in other regions of the Philippines.[13]

At a more local scale, fishers usually equate 'true poverty' (*mahirap na mahirap*), with an extremely basic level of subsistence only. Members of these households are not able to make any investments; have no capital or savings of any sort and frequently live through debt or from day to day on what they are able to sell or catch from the sea. Some fishers of Esperanza regard themselves as belonging to this category. When I asked one fisher, Don, what he thought about poverty and his status, he stated:

> You see my house. [He lived in a nipa hut with no appliances, no electricity, and few material possessions.] I'm extremely poor. The highest income I get is ₱2000 [about US$40] in one month if I'm lucky. Some months I earn absolutely nothing and I have to go into debt. If there is a storm, I have no backup, no other option. I just wait around. I have the life of the ocean … my life is in God's hands. But that's life isn't it?

Most of the fishers however, were able to state they felt they were merely 'poor' or 'average', in that they were able to save limited amounts of money at different times, and make various investments, albeit small-scale. Many fishing households explicitly expressed the goal for their children to gain a good education and thus move out of fishing. For many of these households, fishing had been an establishment strategy for the family over multiple generations, as the quote from Eder (2003: 215–6) earlier described—although the parents often continued to fish throughout their working lives, their goal was to establish themselves enough so that their children would not have to fish as well. One young man, Ricardo Calvino, was somewhat of an exception in deciding to pursue fishing as a career. His mother, Melinda was bemused at her son's insistence:

> We were able to save up enough money through fishing to send Ricardo to finish high school and get a college degree so he didn't have to fish for a living. Our [referring to herself and her husband Ronny, the captain of

---

13   A few residents did leave to seek domestic work in Japan or construction work in Korea.

a commercial fresh grouper boat] hard work and success in fishing has made it possible for our children to go to college. But Ricardo decided not to go on with his education, he really likes the sea. So now he is back with Ronny, fishing. It's strange, isn't it? So we told him that if he wanted to keep fishing, he had to be a leader and eventually take over from Ronny as captain.

Sometimes these discussions of poverty and status take place within the context of a discussion of a household plan or economic strategy (*plano*) (Eder 2006). For the majority of the households I interviewed, their household plan was fairly simple: 'just until there' (*hanggang diyan na lang*). By this, they meant merely increasing their incomes slightly, through the development of their fishing activities by buying a better boat, transferring to a better fishery, and so on. For most fishers in Esperanza, their conceptions of potential economic strategies are greatly constrained by the lack of alternative economic opportunities. As such, remaining as a 'legal fisher' entails a life of relative poverty.

Wright pointed out the prevalence of such a view some time ago:

> Knowledgeable residents decry the lot of the man employed on a local fishing boat more than that of the *'kainginero'* [someone who practices *kaingin*, or shifting cultivation]. At least, they say the *kaingin* farmer has a measure of independence, gets something at the harvest, and can do other things on the side in the meantime. The *pescador* (fisherman) is a slave to his vessel, receiving his pitiful return only after the owner takes his profit and the expenses of the boat are met (Wright 1978: 124).

While not all fishers describe themselves as truly poor, they consistently represent themselves as belonging to a humble occupation. Many fishers described themselves as 'small people' (*mga maliit*). Even those fishermen who owned commercial fishing boats and were relatively well off by the standards of Esperanza would represent themselves (to foreigners like myself, to those working in Coron town or to those in Manila) as 'just a fisher', implying that they were poor and low in status. Importantly then, despite the range of diverse livelihood practices and levels of success described in Chapter 3, the rhetoric of the poor moral fisher is shared by fishers of all types.[14]

Closely linked to such representations is the perception that equates legal fishing and the experience of being poor with family. Bong, for example, said that 'the main value in my life is my family. All my work is focused on helping my family have a better life. I have a good family so I'm very content. You see all those people in town, the rich people, they forget about their family so even though

---

14 This is reminiscent of Peluso's (1992) description of villagers uniting across class differences in an act of resistance, a theme I return to in the conclusion chapter of this book.

they're rich they're not as content'. Similarly, fusilier-fishing captain Mark told me that 'even though we are a poor fishing family, we are happy. We are a happy family because we love each other, not because we are rich'. The quote by Carlos Marquez earlier in this chapter also makes a similar point, where he declares his opposition to illegal fishing in the context of caring for family and children. Here, fishing becomes solely about supporting family. In this way, legal fishing, and the poverty that goes with it, becomes moral through an identification with family—the absolute cornerstone and centre of moral life in the Philippines (Mulder 1997, 2000).

I have argued that fishing is seen as a difficult job; one that is frequently viewed as an establishment strategy or job of 'last-resort' as opposed to an ordinary career; it is viewed as a job of extremely low social status closely linked to financial poverty and uncertainty; and to family. Thus, legal fishing has come to symbolise the poor and humble Filipino who is so valorised in Filipino popular culture through media representations, film and events like the People Power Revolution of 1986. This valorisation relates to specifically Filipino discourses about pity, and the rights of poor people to live their lives with security and dignity. It has important implications for how fishers try to promote their livelihood practices in the face of attempts by people more powerful to curtail their livelihoods, as I detail through the rest of the book.

Conclusion

Fishers in Esperanza represent their practices as: harmless to the environment; legal; moral; and closely tied to poverty. Evidently, these points are valid, and I am certainly not dismissing such representations as false claims or untruths. Illegal fishing is extremely destructive to the marine environment and fishers are among the poorer sectors in Philippine society. What I have argued in this chapter, however, is that the discourse of fishers about these issues develops and exaggerates these two points, infusing them with political and cultural ideas about social relations, morality and the meanings of poverty.

This discourse of the poor moral fisher links ideas about the environment and poverty, and so I argue it is one way of understanding the poverty-environment relationship. For fishers in Esperanza, legal fishing equates with a negligible impact on the environment and a close association with poverty. It is also seen as moral, partly because of its lack of environmental impact, but also because of these links to poverty and family. Illegal fishing on the other hand, is associated with ill-gotten wealth and great damage to the environment. From this perspective, poverty is equivalent to legality and morality, whereas environmental damage is linked with immorality and wealth. Importantly, it follows that any external interventions or regulations should focus on restricting the capacity of those who degrade the environment, while simultaneously bolstering the capacity of harmless, legal fishers to pursue their livelihood.

I have therefore tried to demonstrate how the relationship between poverty and the environment in the Calamianes must be considered from multiple perspectives, as I outlined in the first chapter. Local residents understand and express the relationship as a political one. This political nature of the discourse is reflected both in the way it highlights the injustice and immorality of the relationship, and, as I shall argue in later chapters, the strategic way in which it is deployed against conservation regulations. I have also tried to illustrate in this chapter how such claims of fishers are embedded within particular cultural understandings of poverty and morality. In the Calamianes, the relationship between poverty and the environment can be seen to be as much reflective of local discourse as of any empirical reality. The rest of the book goes on to argue how this shapes and influences the marine resource management process. Chapters 5–7 examine various aspects of this discourse related to marine resource regulation. I begin by focusing on how fishers have responded to the introduction of marine protected areas.

5. Fishing, Dive Tourism and Marine Protected Areas

One significant effort in regulating coastal areas of the Calamianes has been through the creation of a series of community-based marine protected areas (MPAs). These MPAs, established by a number of organisations and institutions over several years in particular diving sites, included a system of user fees. Fishers' responses to these MPAs were ambiguous. I argue that these responses can be best analysed with reference to the ideas and representations bound up in the discourse of the poor moral fisher. I demonstrate how the attitudes and behaviour of fishers went a great way to shaping the particular character of these MPAs.

Essentially, support for MPAs among fishers in the Calamianes was closely tied to the question of whether or not they would have any impact on their livelihoods. This varied within and between communities, but overall fishers favoured the development of MPAs that were strongly directed towards providing benefits to local communities. MPAs were understood here as interventions that should not impinge on the activities of poorer small-scale fishers; it was argued that such interventions should support those activities. Local government officials, overall, supported these claims of fishers, seeing in MPAs a chance to bolster their conservationist credentials as well as providing extra revenue to the municipal treasury. For those in the dive industry, strongly affected by the transfer of management to communities and the imposition of user fees, responses were extremely negative. They argued that the MPAs were not about conservation at all; this was just a sham hiding the real desire among local government officials and communities to make money out of the dive industry.

I begin this chapter by reviewing some background related to the history of fisheries and coastal management in the Philippines. After an introduction to the scientific bases of MPAs is an analysis of some of the literature focused on the management of MPAs and their links with dive tourism. The majority of the chapter examines how local fishers, tourism operators, and local government officials responded to the creation of MPAs in the Calamianes. I conclude with a discussion of relevant policy implications that my analysis holds for the nexus of MPAs, fishing and dive tourism.

Background to Coastal Resource Management and the Development of Marine Protected Areas

Fisheries and Coastal Resource Management in the Philippines

Coastal resource management, as detailed in the literature, has not existed for most of the history of the Philippines.[1] The story of fisheries in the Philippines is perhaps familiar to those with knowledge of fisheries in other parts of the world (see for example Roberts 2007). As Butcher (2004: 193) describes it, for a certain period 'catches increased enormously as fishers based in the Philippines extended capture into ever deeper and more distant waters by applying new techniques and new versions of old methods and by doing so at a rate faster than that at which fish populations were depleted and ecosystems disrupted'. Technologies changed from fish traps and corrals in the early part of the twentieth century, through to the use of bag-netting (basnigan), which grew in number in the post-war period, through to the prevalence by the 1970s of the trawler (particularly the otter trawl) and of purse seining vessels.

Since this period, scientists have become more aware of just how much damage has been done to the marine environment of the Philippines. In 2004, the Bureau of Fisheries and Aquatic Resources (BFAR) produced a comprehensive assessment of the state of Philippine marine fisheries. Providing a snapshot of the problems and issues faced by the fisheries, it emphasised eight defining characteristics:

- depleted fishery resources;
- degraded coastal environment and critical fisheries habitats;
- low catches, incomes and dissipated resource rents;
- physical losses and/or reduced value of catches due to improper post-harvest practices and inefficient marketing;
- inequitable distribution of benefits from resource use;
- intersectoral and intrasectoral conflicts;
- poverty among small-scale fishers; and
- inadequate systems and structures to manage fisheries (BFAR 2004: 345).

As White and others point out (Courtney and White 2000; White et al. 2002), in more recent times two major forces have influenced the development of

---

1   Butcher (2004) and Spoehr (1980) have both provided detailed and comprehensive accounts of the history of fisheries and fishing technology in the Philippines.

coastal resource management in the Philippines. First was the implementation of various projects and programs dealing with coastal resource management, funded by a range of national and international governments, NGOs and as well as other organisations. Major projects such as the Central Visayas Regional Project, the National Integrated Protected Area Project, the Coastal Resource Management Project and the Fisheries Resource Management Project were all multi-sited attempts at establishing a range of coastal resource management measures, including MPAs, community-based coastal resource assessments, mangrove reforestation, mariculture, strengthening fisheries law enforcement, and organising fisheries associations (White et al. 2002).

The second major force was the devolution of authority over coastal resource management from national to municipal, city and provincial governments, as manifested in a range of legislation. Most important were the Local Government Code (LGC) of 1991 (*Republic Act 7160*), and the Fisheries Code of 1998 (*Republic Act 8550*). The LGC was a wide-ranging reform that decentralised a range of government functions and powers to the municipal, city and provincial levels. In the coastal resource management sector, the LGC defined municipal waters as extending 15 km offshore, and gave municipal and city governments (local government units, or LGUs) authority over and a mandate to manage these waters. Based on this provision, the operation of commercial fishing vessels (defined as over three gross tons) became illegal within these municipal waters without a permit. The Fisheries Code of 1998 strengthened these provisions, specified the roles of LGUs with regard to management, and provided more detail on various prohibited acts and penalties (ELAC 2004).

Integrated Coastal Management or ICM [2] has emerged partly from the international debates and trends prevalent in the conservation movement during the 1980s. The focus on decentralised management and increased participation were a part of the broad trend towards community-based natural resource management and integrated conservation and development projects that occurred throughout the world (McShane and Wells 2004; Brosius et al. 2005). More recently, ICM has aimed to incorporate aspects of the move towards ecosystem-based management as well, and the FISH project, which has one site in the Calamianes, has been cited as the first example of ecosystem-based management in the tropics (Christie et al. 2007).

The goals of all such ICM projects are wide-ranging, according to White et al. (2005). They aim to: improve biophysical conditions; increase the level of stakeholder participation in the decision-making process; contribute

---

2 Many of the coastal resource management projects in the Philippines use a variety of terms to characterise their approach, including co-management, community-based management and more recently ecosystem-based management. I use the term integrated coastal management (ICM) here, as this is the most general term that encompasses a wide variety of more specific approaches.

to economic returns and livelihoods; strengthen legal and policy frameworks; strengthen the capacity for law enforcement; and build durable institutions beyond leadership changes. Christie et al. (2005: 471) state that 'ICM represents an appropriate middle ground between those advocating mainly for social and economic justice and those advocating mainly for environmental preservation'.

Christie argues that ICM is characterised by two key features: balancing conservation and development and ensuring multi-sectoral planning, and increasing levels of participation among various levels of government and other stakeholders (Christie 2005a: 209). The tools adopted in ICM projects include various measures such as those mentioned earlier; however, MPAs in particular tend to form a significant component of any ICM regime (White et al. 2005: 274). In 1974, the first MPA was established in Sumilon Island in Cebu, by researchers from nearby Siliman University in Dumaguete. Sumilon, and nearby Apo Island in Negros Oriental were well managed and well documented, and Apo Island in particular has served as a 'model' for demonstrating the validity and effectiveness of MPAs throughout the world (Alcala and Russ 2006). Since the establishment of these MPAs, '[a]s of 2008 at least 985 MPAs had been established in the Philippines', and 'they covered approximately 14,943 km2' (Weeks et al. 2010: 533).

The Scientific Bases of Marine Protected Areas

The general scientific purpose of MPAs is simple. By restricting fishing access to one particular area, the goal is that fish will be able to reproduce in peace. From this protected area, the 'spillover effect' is anticipated to occur (Russ 2002). Here, the fish that are born within the protected area spill over into the waters surrounding the MPA, therefore increasing the number of fish available for fishers to catch. In theory then, MPAs are widely seen as organised spaces that can satisfy various stakeholders. For conservationists and SCUBA divers, the MPA provides a pristine, untouched habitat where divers can admire the fish and reefs. For those more interested in more general fisheries management, and for the fishers themselves, the MPAs potentially offer an increase in the fish stock *outside* the MPA through the spillover effect. The hope with MPAs then is that while different stakeholders such as local governments, conservationists, dive operators, and fishers may have different motivations for creating MPAs, their end goal should be the same and so the interests of all stakeholders will be satisfied (Roberts and Hawkins 2003). Therefore, in theory, MPAs are different to terrestrial protected areas in that instead of offering a straight payment (in the form of ecotourism or alternative livelihoods) for giving up rights to the resource, the goal is that the MPA *itself* will provide more fishing income in the areas outside, through the spillover effect.

There are some significant concerns with the spillover theory however. As Sale (2002) points out, the science showing that protection works within MPAs is actually much stronger than the science showing that fish stocks increase in the area outside the MPA. He argues that there is 'no evidence that MPAs serve to enhance fishery yields in the region surrounding them to a degree that fully compensates for the loss of fishery access to the area they enclose' (ibid.: 367). Hilborn et al. (2004) also point out some of the limitations of MPAs, noting that they will not work for many other more mobile species, and that the evidence of any positive effects for fisheries more broadly is scarce. Fisheries effects take a long time to occur, and mostly only via adult spillover within a short distance from the edges of the reserve. However, as MPAs 'can provide conservation benefits irrespective of any fishery management benefit they offer' (Sale 2002: 367), the construction of many MPAs indicates a tendency for the interests of conservation to be prioritised over those of fisheries management.

Links between Marine Protected Areas and Dive Tourism

There is a considerable literature dealing with the ecological bases, fisheries science, and legislation relating to MPAs in the Philippines (for example Pollnac et al. 2001a; Christie et al. 2002; Christie 2005b; Alcala and Russ 2006). In particular, a subset of this literature has addressed the potential relationship between dive tourism and MPAs (Arin and Kramer 2002; Oracion et al. 2005; Depondt and Green 2006; Majanen 2007). In one exploratory study, Arin and Kramer (2002) found that SCUBA divers in the Philippines would be willing to pay a user fee of US$4 in selected marine sanctuaries. They suggest that by implementing a system of user fees in areas of high dive tourism, potential annual revenue could reach up to US$1 million in particular diving areas. This revenue, they argue, could support the maintenance of MPAs and provide alternative livelihood opportunities for fishers impacted by the MPAs. Depondt and Green (2006) also point to the potential to raise income through user fees, noting the high level of dive tourism in Southeast Asia and the fact that many of these dives take place within MPAs. They argue that the 'potential of diving user fees to address important funding problems of MPAs is present but not sufficiently exploited' (ibid.: 201). However, they caution that 'the importance of transparency in how the revenue is used is central' (ibid.) to the acceptance among dive operators of user fees, noting the high level of suspicion directed towards governments among many dive operators in Southeast Asia.

A key feature of such literature is the attempt to evaluate and discover features and factors that influence the success of MPAs. In an influential article, Pollnac et al. (2001a) declare that six primary factors appear to be fundamentally important to the success of MPAs in the Philippines:

- population size (relatively small);
- perceived crisis in terms of reduced fish populations before the MPA project is started;
- successful alternative income projects;
- relatively high level of community participation in decision making (high on the democracy scale);
- continuing advice from the implementing organisation; and
- inputs from the municipal government.

Success is broadly defined in terms of several factors such as increases in corals and fish stocks, adherence to the MPA rules and the level of community management.

The Importance of Social and Political Factors

The factors Pollnac et al. (2001a) analyse are certainly important to the development of MPAs. However in their focus on identifying commonalities between various MPAs, the authors leave somewhat understated the role of local social and political factors that can also shape outcomes. It is also important to bear in mind the experiences of writers such as Van Helden (2004: 99) in Papua New Guinea, who found MPAs 'turn out not to be established on the basis of ecological best practice but more on a fuzzy practice of "making do" with the available means under the existing social and economic circumstances'. In other words, creating and developing MPAs are not merely technical endeavours but are highly dependent on locally contingent social factors that are not easily modelled.

Therefore, it is important to consider the literature that examines some of the social bases and effects of MPAs. Eder (2005) has pointed out the institutional weakness of ICM in another municipality of Palawan, showing that approaches towards and the impacts of MPAs (as well as other elements of ICM) are differentiated by class, gender and ethnicity. Oracion has written the most comprehensive study of the social nature of MPAs in the Philippines; his Ph.D. thesis (2006) addresses the cultural politics of MPA construction in Negros Oriental province. With similar elements of tension between the fishing and tourism industries, he stresses that MPAs are a 'political and cultural space' just as much as a 'space for conservation' (ibid.: vi). Elsewhere, Oracion et al. (2005) point to the impact of MPAs on the livelihood of fishers. They identified significant tensions between the fishery and tourism sectors in Mabini, Batangas, an area of the Philippines where numerous MPAs have been set up to work with dive tourism. They attributed much of this tension to different perceptions of the purpose of an MPA, arguing that 'unless a common understanding and interpretation of the significance of MPA [sic] is forged among multiple stakeholders its management will continue to be filled with tension and threaten community solidarity' (ibid.: 408). They argue that the tourism and fishing sectors were both interested in MPAs, but that the livelihoods of the tourism sector 'support aesthetic conservation and foster a tourism economy, while those of the [fishing sector] underwrite extractive conservation and a fishing economy'. Ultimately, in Mabini, the tourism sector was able to control the MPA management. Majanen, (2007), working at the same location of the Philippines, found a similar set of tensions in her research. Fishers, according to Majanen, perceived that conservation efforts in the form of MPAs and user fees benefited the tourism sector, while marginalising their livelihoods.

Similarly, MPAs in the Calamianes are inevitably produced and transformed by social and political concerns. My approach here is to focus on how the claims of fishers in particular about poverty, morality and the environment shaped the process of developing MPAs. For fishers, no purely technical, scientific management of MPAs would be fair or legitimate without taking into account their claims. Dive operators, too, felt strongly about their rights to management over the areas covered by the MPAs, but for different reasons. The outcome was a set of debates over conservation and fisheries management that masked the real debates about livelihood and territoriality that lay beneath. Fishers and the dive industry pulled the MPAs in different directions, and the outcomes reflected the ways in which particular groups of people were able to advance their interests successfully. I argue that the perspectives that fishers bring to the creation of MPAs ought to be considered more closely by policymakers, because of the key role they play in shaping the ultimate character of MPAs. In particular, my analysis of these attitudes and practices calls into question the claim that MPAs are able to satisfy all stakeholders.

Marine Protected Areas in the Calamianes

In the Calamianes, numerous reports, such as those relating to the live fish trade (for example, CI 2003; Padilla et al. 2003), have pointed to the growing impact of a lack of management in coastal areas. Because of these local concerns and the broad national trend that I outlined earlier, the push for MPAs and coastal management has increased in recent years. Before 2004, there were several established MPAs, which were mostly organised and run by resorts in the area. But a stronger push began in 2004, when there was a large effort by a Japanese funded project of the provincial government—SEMP-NP, or the Sustainable Environment Management Project of Northern Palawan—to create a series of MPAs that would work with tourism. The goals of SEMP-NP were to establish a mechanism for the collection of environmental conservation fees in selected areas. In developing the model for the user fees in the Calamianes, the planners explicitly engaged with the ideas expressed in recent coastal management literature (Arin and Kramer 2002), which indicated that divers were willing to pay user fees for marine conservation (Green 2004). The role of dive tourism in facilitating marine conservation and user fees became the focus of the project. Simply, the idea was that divers coming to dive on the wrecks and on other selected dive sites would pay user fees. These would fulfill two aims: firstly, provide for the maintenance of the marine parks through payment of guards and wardens and their boat, together with marker buoys, signs and ropes. Secondly, a proportion of the funds were also intended to enable local municipalities to support other marine management projects, as well as provide a local barangay development fund to assist those fishers who would be affected by displacement from their fishing grounds. This money was supposed to be for such remedial measures as the development of alternative livelihood projects or the creation of a credit cooperative. After a series of public hearings throughout 2004 and 2005, the specific municipal and barangay ordinances were enacted at the end of 2005. A user fee of ₱100 (US$2) is now to be paid by each person who dives in the marine parks, and responsibility for management has been devolved to the local coastal communities.

Plate 5-1: Promoting marine protected areas in Coron town.

In 2006–07 there were over a dozen MPAs developed by a range of conservation organisations throughout the Calamianes in various stages of implementation. An important point to note is that many of the MPAs instituted by other organisations were developed for the purposes of conservation and fisheries management—not with the primary goal of engaging with tourism. Their goal was to increase fish stocks, not to provide financial benefits for affected communities. Target sites for these MPAs were selected on mostly ecological grounds that had been determined through extensive marine survey work. At the request of the local communities, however, user fees were incorporated into those MPAs as well. My particular interest in this chapter is to establish how *all* MPAs implemented by various agencies in the Calamianes have been commonly interpreted, understood and adapted to by local fishers.

Fishers and Marine Protected Areas

The understandings of fishers about the relationship between fishing, morality and poverty discussed in earlier chapters shaped the ideas fishers expressed about the potential regulation of small-scale fisheries through MPAs. In various meetings, informal conversations and through their fishing practices, fishers would fall back on these understandings as a way of legitimating, defending and advancing their interests. I now demonstrate how these understandings would sometimes support the creation of MPAs, and sometimes undermine them. In both instances however, the themes of the poor moral fisher—the lack of responsibility for environmental degradation and the need to address poverty among fishers—were strongly expressed.

I was living in Esperanza during the period of MPA establishment in the Calamianes. I was not present at any of the meetings that were held at Esperanza during the original planning and implementation of the marine park there (2004-05), but I learned a great deal about attitudes towards MPAs through interviews and informal conversations throughout the course of my ongoing research. I visited the neighbouring community of San Andres regularly during 2006–07, and was present for most of the meetings regarding the development of the MPA there. Data about other MPAs in the Calamianes (including those focused on the wrecks) I collected while on short visits to these sites, and interview data from Coron town.

Esperanza Marine Park: Different Impacts, Different Responses

The example of the Esperanza MPA serves as an introduction to show how fishers' responses to MPAs were shaped primarily by the impact they had on their livelihoods. The Esperanza MPA was an initiative of the SEMP-NP, which included user fees to the snorkelers and divers visiting a popular reef site. Throughout 2004 and 2005, discussions held with community members of Esperanza, dive operators and municipal officials, resulted in the municipal council passing the ordinance to legislate the formation of the Esperanza Marine Park towards the end of 2005. The MPA covers a total area of fifty-two hectares. Fifteen hectares of this is composed of a strict protection or 'no-take' zone, where fishing is prohibited. The location of this zone was calculated from observations made by scientists working on the project, and adjusted for the presence of a large fish cage owned by a resident. The other 37 ha is a 'buffer zone' where certain regulated forms of fishing are allowed, such as the use of hook-and-line. The Esperanza MPA brought in more than ₱150 000 (US$3000) during 2006 from user fees. Under the ordinance, 45 per cent of this was supposed to go directly back into the project for the provision of necessary items such as marker buoys, ropes, and a boat for the fee collectors. Twenty-five per cent was potentially earmarked to go back into the community in the form of a credit cooperative, through a barangay development fund. The remaining 30 per cent was supposed to return to the municipality for the purposes of future coastal resource management projects. There was ongoing tension over this proposed distribution of payments throughout 2006–07. Unlike many other MPAs in the Calamianes, and indeed the Philippines (White et al. 2002), the Esperanza MPA was extremely well enforced, despite some informal resistance that arose during the year. The strong support of the barangay captain to enforce the MPA and the relatively close location to Coron town (the Coastguard and Police base) were the two key reasons for this.

A survey conducted by those implementing the MPA at Esperanza in 2004 referred to 90 per cent of the community being in favour of the project (Green 2004: 11). It cited the fact that the majority of fishers in Esperanza did not regularly use the reefs within the proposed MPA as the central factor behind this, pointing out that commercial fishing took place much further away. Two years on, during my fieldwork in 2006–07, many of the fishers that supported the MPA notionally still did so, in part presumably because their fishing activities took place well outside the protected zone. In the past, the area had allegedly been a popular haunt for fishers using cyanide, and so some residents believed having the MPA would be a good opportunity to keep these fishers away. Most others simply thought that the user fees would be valuable to the community, or that the presence of tourists would be a good opportunity for

Esperanza to develop its own businesses such as guesthouses or souvenir shops. Some residents already participated in tourism-related industries such as boat carpentry and boat driving, and they saw the creation of the Esperanza MPA as facilitating the growth of these opportunities.

Similar to the MPA that Eder (2005: 159) analysed in San Vicente, however, the MPA at Esperanza negatively affected the poorest fishers. In both San Vicente and Esperanza, the MPA was located close to shore. Because of this, those fishers with pump boats or access to them (or the larger commercial boats as well in the case of Esperanza) were not disadvantaged by the creation of the MPA. Instead, it was only those poorer fishers who paddled out to do their fishing who had to travel further to fish in different grounds.

These fishers grumbled against the existence of the MPA during 2006–07. When I asked one such fisher where his usual fishing grounds were, for example, he responded in a manner typical of many of these affected fishers: 'Usually I used to fish with hook-and-line just here, you see, just here [points to the area covered by the MPA]. Now, things are harder though, ever since they introduced this … "Marine Park", laying emphasis in English on 'Marine Park'. When I asked him what he meant by his seemingly sarcastic emphasis on these words, he shrugged uncomfortably before saying that he did not understand why the government was trying to introduce things like this that would hurt people like him. 'I am just a poor fisherman … the government should be helping people like us, instead of making laws that will hurt us.'

Anecdotal evidence among these fishers and NGO workers suggests that the process of community consultation with this sector had been inadequate, and their voices insufficiently heard.[3] Some other fishermen, who did not previously fish in the MPA area but were sympathetic to those fishermen who did, were also privately critical. Community members who opposed or had reservations about the MPA could not understand the fisheries and conservation logic underlying it: 'This Marine Park just causes problems for *mga maliit* [the small people] … if they are trying to make the fishing better they should be concentrating on arresting the illegal fishers instead', said Miguel, a net fisher. Similarly, Eddie (another net fisher) argued that 'this Marine Park will just be good for the tourists. They are allowed to go and dive there; why aren't we allowed to fish there anymore?'

---

3   See Eder 2005 for a similar account of the ambiguous nature of 'participation'. Li (2007: 192–229) also offers an insightful critique of how participation in development projects often proceeds towards pre-defined outcomes.

Other residents were sceptical about the potential benefits of tourism:

> The benefits of tourism will only go to the LGU. It will bypass all the
> fishers. It's fine if you have a relative in the LGU, but most fishers don't
> have that of course. We are poor people here. If tourism keeps growing,
> our livelihood will suffer. You come back in five years time and see what
> will have happened: banned, banned, banned!

Following on from their representation that only illegal fishers cause damage
to the environment, such residents argued that the creation of the MPA was
ignoring the real problem of illegal fishing and simply punishing the small-scale
fisher. In Chapter 7, I detail how these perceptions of MPAs are also strongly
linked to perceptions of poor and corrupt governance. Here however, I wish
to emphasise how opposition to MPAs was also based on the arguments that
I outlined in Chapter 4: that their fishing activities were completely harmless
to the environment, and that their poverty ought to be deserving of assistance
from the government. Unfortunately for these fishers in Esperanza, however,
they were in the minority.

Responses to the Esperanza MPA demonstrate the way in which support for
an MPA is determined greatly by whether a fisher stands to lose or benefit
economically by its creation. As only a minority stood to lose by the creation
of the MPA (the poorer hook-and-line fishers), the MPA was mostly initially
supported, despite the reservations of some of the other fishers. As I describe
later in this section however, the level of enforcement in Esperanza was
significantly stronger than in other areas of the Calamianes. This was a result
of contingent factors shaped by the actions and values of locally politically
influential characters, and the fact that the MPA was located close to town.
While in this case the MPA was implemented and enforcement was maintained,
muted complaints continued throughout 2006–07 among those fishers who had
been affected. In the next example at San Andres, more fishers were affected by
the MPA, and so the discourse of the poor moral fisher was able to achieve more
traction.

Other Marine Protected Areas: Supporting Fishing Practices and Taxing Tourism

Meetings held during the development of other MPAs in the Calamianes show
further the ways in which fishers constructed ideas about poverty, morality and
fishing, and how they actually shaped outcomes. In these examples, I show how
fishers viewed the development of an MPA as a way of constructing a form of
marine tenure, or as a way of protecting their own resource use practices. As I
introduced earlier, control over waters extending 15 km from the shoreline rests

with the municipality, and only artisanal or municipally registered commercial vessels are (ostensibly) allowed to fish within this area. Outside of this zone, the sea is open to any commercial vessel. Importantly then, there are no barangay or sitio waters under the law. According to numerous NGO workers and government officials who have worked on ICM projects, the barangay will usually have in practice some level of control of waters immediately near its particular land, as is also indicated by the example in Chapter 4 where the baby purse seine boats negotiate with the barangay to fish there. However, the municipality is the smallest political unit in the Philippines that can formally exercise jurisdiction and legislation over marine territory.[4]

I attended several meetings concerning the development of an MPA at San Andres, a neighbouring sitio to Esperanza. Government officials from the municipal Department of Agriculture and an NGO provided technical assistance for this MPA. Unlike the Visayan migrants at Esperanza, the community at San Andres is composed mostly of Indigenous Tagbanuas. Here, farming is the primary livelihood, and fishing is not as prominent as in Esperanza. While some San Andres residents crew on the Esperanza fusilier boats during habagat, there are no commercial boats based in San Andres. The relative ease of farming in San Andres compared to Esperanza is made possible by the presence of a river, which facilitates irrigation, and the availability of secure land tenure.

A key feature of the proposed MPA was to include user fees, which residents demanded as an essential part of the process. During a later meeting discussing how the MPA would benefit local residents, one community leader enthusiastically exhorted the residents of San Andres to get behind the project:

> This can be our project ... together we can show everyone that even though we are Tagbanuas, we can manage this place and we have a right to declare who comes into the place, no matter who they are! Tourists will have to ask permission to visit this marine park, and they will have to pay our community to do so!

Explicit in this declaration was that one of the prime roles of the MPA was to 'empower' the community to address some of the massive inequalities between the coastal residents and other more powerful or wealthy figures such as foreign tourists.

Following the standard process for implementation, the San Andres MPA was to be divided into different zones defined by what activities could be allowed

---

4   Recognition of the rights held by Indigenous Tagbanuas to the marine territories surrounding Coron Island (PAFID 2000) made this an exception. However, forms of customary marine tenure such as these and others found in other parts of the Asia-Pacific region (for example Hviding 1996) are relatively uncommon in most parts of the Philippines.

inside. During the first round of community meetings, San Andres residents and NGO workers firstly discussed this point, drawing a large map of where the core zone could be located and what activities could be allowed inside the buffer zone. The core zone was agreed upon relatively easily. This was based primarily upon the manta tow observations that had taken place several weeks earlier by NGO staff, but was adjusted to allow for the presence of several seaweed farms and fish corrals (a type of fish trap) owned by San Andres residents that lay within the proposed core zone. Discussing what activities were to take place within the (much larger) buffer zone was a little more complicated, because of the wide variety of techniques that residents normally used within this zone. It was clear from the animated discussion that residents did not want to give up rights to fish in this area. One by one, San Andres residents spoke about the various fishing gears used. So, by the end of the day, it was provisionally agreed that hook-and-line fishing, crab pots, seahorse fishing, net fishing for the marine ornamental trade, and a specific type of spear fishing were all to be allowed.

Other sorts of fishing activities were banned within the buffer zone. Prominent among these activities was the use of 3-ply, a particular type of gillnet that was widely used among fishers of Esperanza. My companion at this meeting, a community leader from Esperanza, quietly remarked to me at the end of the first meeting that he was going to have a big problem on his hands—virtually all of the 3-ply fishermen from Esperanza used this particular area near San Andres.

The next day another meeting occurred where more details of the proposed MPA were worked out. This time, residents from Esperanza turned up to defend the use of 3-ply in the buffer zone. The main spokesman for the San Andres residents, an older landowner named Ferdinand, argued that 3-ply should be banned from use within the buffer zone. He stated that it was an extremely efficient gear, and that if it was not banned then there would be no point in having an MPA at all. Officers from the Department of Agriculture and the NGO who were providing technical assistance agreed with this interpretation. However, they also were committed to their role as providers of technical assistance only: 'We just give them the information, and then they decide what to do' was how one NGO worker at the meeting characterised the process of community-based coastal resource management.

Following much discussion, one fisher from Esperanza got up and made an impassioned defence of the use of 3-ply. He described how 3-ply had been his livelihood for a long time, since 1989, and that there were still plenty of rabbitfish (the targeted species of 3-ply).[5] Cyanide, he emphasised, was the real problem—not 3-ply fishing. The area in the proposed MPA opposite San Andres

---

5   Rabbitfish are a resilient group of fishes that reproduce quickly (FISHBASE 2008).

had been his primary fishing ground for many years, and if he was stopped from fishing there he would have to travel much further to catch enough fish. He was poor, he declared; the creation of this MPA would needlessly bring extra hardship on his family. He pointed out that many of the residents of San Andres had access to alternative incomes through their land ownership and farming, whereas those living in Esperanza were forced to rely virtually exclusively on the sea for their livelihoods:

> People here, these people have farms, people can grow cashews, rice, bananas. But, for us in Esperanza, we have to fish everyday! Me, for example, I don't have any land at all, everything I earn is from the sea. This is what I rely on to survive—I am not a rich fisherman, no, I am just a poor fisherman and what I catch is just to survive.

This powerful plea thus explicitly invoked the 'right to survive' (Blanc-Szanton 1972: 129). Here, the underlying notion was that no matter how great the need to reduce fishing pressure around the area, this should not affect his fishing practices. As it was, his net fishing brought him enough income 'just to survive'; any impact on this income would make his life even more difficult. From this view, those in charge of regulating marine resource use had an obligation to ensure that his 'right to survive' was not compromised.

During the open discussion there was a very strong emphasis on speaking politely, and maintaining the proper ways of addressing each other through 'smooth interpersonal relations' (Lynch 1970), but after the meeting concluded a great deal of privately expressed tension was evident. The fishers from Esperanza asserted that Ferdinand was trying to obtain the benefits of the MPA solely for the residents of San Andres. They resented the fact that various techniques such as spear fishing and seahorse fishing were to be allowed within the buffer zone just because these were techniques used by San Andres residents, whereas 3-ply was to be banned because there were very few San Andres residents who used this technique. (The one 3-ply fisher from San Andres who attended the meetings was too afraid to speak up because he had utang na loob, or considerable personal debt, towards Ferdinand). They angrily muttered that the San Andres residents had no legal right to exclude them from the waters near San Andres: 'There is no such thing as *barangay*[6] waters, there is only municipal waters!' They also declared that the Tagbanua people were selfish, citing the common perception among other Filipinos in Coron that 'it is their mentality to take what they can get without doing any work for it'.[7] Similarly, Ferdinand and other residents who supported the banning of 3-ply in the sustainable use

---

6 Because of its relative isolation, residents often refer to San Andres as a separate barangay even though it is in fact a *sitio* within the same barangay as Esperanza.

7 As I describe later in this chapter, this was the same charge levelled at all Filipino fishers (not just Indigenous Tagbanua) by some foreign tourism operators.

zone were incensed at the behaviour of the 3-ply fishers from Esperanza: 'Can you believe those guys?! They just came here to make sure they could keep on using 3-ply without getting involved in the rest of the project! *May high blood ako*! [I have high blood pressure!]',[8] exclaimed one of the government officials to me after the meeting.

Both groups were clearly acting in their own interests by trying to protect their exploitation patterns, and neither group was seriously interested in regulating their own practices. For the San Andres residents, the MPA seemed to them to offer a way of legitimating their own practices of marine resource use, and excluding the practices of all other fishers. When the Esperanza fishers declared that 'there is no such thing as *barangay* waters, there is only municipal waters', they were in effect resenting the way in which Ferdinand and other residents had attempted to construct an artificial form of marine tenure for the sitio. Esperanza residents opposed the creation of the MPA because it was seen as limiting their patterns of resource use for no good reason. For all of the fishers, the MPA was viewed as something that should support the community, something that was solely about supporting their livelihoods: such MPAs should not impact on existing fishing practices, or do so only minimally, and these MPAs should be accompanied by strong benefits derived from user fees, which should return directly to the communities.

In the end, the debate reached a stalemate, and as of 2010 the MPA is still a 'paper park'. In effect, 3-ply fishing ended up being recognised as a legitimate form of fishing, and the fishers from Esperanza were seen as legitimate claimants. Most likely, this is because they lived nearby and dealt with people from San Andres on a frequent basis: this points again to the importance of maintaining reciprocal social relationships. The way in which the discourse of the poor moral fisher in this instance received a favourable audience indicates that the likelihood of success when 'appealing for pity' depends at least in part on establishing and maintaining ongoing social relationships.

Examples from other MPAs in the Calamianes conform to the practice of using MPAs to support local fishing practices. In Langko, an isolated, island barangay, one MPA had been developed that encircled one of the smaller islands, Dera. A dispute had been taking place over the ownership rights of Dera for several years. A wealthy family from Coron town claimed to have bought the island in the 1970s from the indigenous Tagbanua landowners in order to harvest pebbles from the beach. Subsequently, they sold Dera to a Realty Corporation, which then came and tried to auction it. The local Tagbanua residents of Langko objected, saying the island was never sold. They claimed hereditary rights over it, as a part of their broader Ancestral Domain Claim throughout that region

---

8   This is a common Taglish expression in the Philippines indicating anger or tension.

of the Calamianes (this claim was still in dispute during 2006–07). So, when a national NGO arrived with the intention of creating several MPAs, local residents strongly supported the idea of creating an MPA around Dera Island. In this MPA, hook-and-line, spearfishing and squid jigs—the primary fishing techniques used by the resident Tagbanuas on the island—were allowed to be used within the MPA, and other fishing techniques not used by the residents (such as 3-ply) were banned.[9] User fees were introduced here as well, as one NGO worker explained: 'The idea with imposing user fees is to let the communities control tourism, not let tourism control them'. Like other conservationist NGOs operating in the Philippines, this NGO is explicit about the need for social justice to be incorporated into conservation. Indeed, one of these NGO workers told me that MPAs such as these were meant to be a 'bargaining chip for local communities to use against more powerful outsiders'. The MPA at Dera can definitely be viewed as a way of protecting local Tagbanua rights over the waters and the island itself.

A final brief example of how the presence of MPAs can be used to exclude outsiders is from Bedalo, another small island barangay. At Bedalo, an MPA had been proposed at one particular end of the island, but by 2007 the negotiations had not been completed and the necessary ordinances had not been created. When 3-ply fishers from Esperanza went to fish near Bedalo, residents of Bedalo frequently accused them of poaching and fishing illegally within an MPA. This was despite the MPA not having been legislated yet, and the 3-ply fishing occurring on the other end of the island from where the proposed MPA actually was. In this instance, unlike in San Andres, Esperanza fishers did not have much contact with residents of Bedalo, and it is unlikely that they had managed to form many social relationships of sufficient depth for the discourse of the poor moral fisher to find a sympathetic audience and 'succeed'. As indicated earlier, fishing in inshore areas close to land (unlike live grouper fishing conducted on reefs that were usually located further offshore) usually required some sort of permission of the local barangay or sitio authorities by outsiders. The development of MPAs serves as an opportunity for local fishermen to strengthen boundaries against these outsiders.

In 2006–07, most of the MPAs in the Calamianes (other than the Esperanza Marine Park) had a very low level of enforcement; many were to a significant degree simply 'paper parks'. When approaching the MPAs based around the shipwrecks, for example, dive operators would frequently report that they had seen fishers fishing on top of the wrecks before approaching the dive boats to extract their user fees. Other MPAs had some level of enforcement during

---

9   During 2006–07, negotiations were underway for a second MPA in Langko; this MPA included regulations on local fishers.

the day, but were unable to continue this during the night. The low level of enforcement of local fishers suggests that they felt that they still had a right to fish there.

The development of MPAs in the Calamianes therefore shows how fishers use the discourse of the poor moral fisher to assert various claims. When fishers do support MPAs, they use them as a means to increase or maintain their own levels of exploitation while trying to gain benefits out of the tourism industry. The rationale behind this support for these MPAs has less to do with conservation or even fisheries management per se, but more to do with the view that: 1) small-scale fishing of the type practiced by fishers in the areas covered by MPAs is environmentally harmless and as such should not be subject to regulation (as opposed to cyanide); and 2) these fishers are extremely poor, and to reduce their ability to fish would be to impinge upon their 'right to survive'. From this perspective, rules for MPAs should not be based on abstract principles of scientific management, but on principles of social equity. Therefore, when support comes it aims to minimise impacts on fishers, and maximise the benefits of tourism. I now turn to a discussion of how tourism operators received these claims, and countered them with claims of their own.

Responses to Fishers' Claims

There were three diving operators in Coron town during 2006–07, and at least half a dozen others based on isolated resorts in different locations of the Calamianes. More operators arrived between 2007 and 2010. In 2006–07, all of the owners and dive instructors (although not all of the divemasters) were foreigners from Western countries. There were various other tourism operators from Western countries as well, including hotel/guesthouse operators and restaurateurs. The creation of the MPAs was cause for a great deal of debate, argument and resentment during this period of my fieldwork research. I focus now on the grievances of the tourism operators and how such grievances are related to perceptions of the Philippines and its people, and perceptions of marine ownership and territoriality.

Tourism Operators: Marine Protected Areas as a 'Money Grab'

The most commonly articulated complaint by tourism operators in Coron was that the MPAs were quite simply not about conservation at all, but about getting money for the local communities and the local government. Essentially, MPAs legitimated what they viewed as a quest among local communities for 'easy money'. Various quotes made by members of the tourism industry make it clear that they resented the claims by communities for benefits out of the MPAs.

I call them 'collecting stations' because they are just about making money.

They are just trying to make money out of the dive industry.

They're just a money grab.

They have a carrier pigeon now; it flies around telling all the communities about how to get easy money.

They cited other examples from the region where local communities had been given control over the management of natural resources, such as when the indigenous Tagbanua community had been granted Ancestral Domain over Coron Island. There, they pointed out, rubbish had accumulated on the various beaches tourists frequented, and the facilities at these beaches and other tourist destinations on the island such as the renowned Kayangan Lake had fallen into disrepair through lack of care and maintenance. Their complaint was that once the communities were given control of the dive sites and MPAs, a similar situation would occur. For example, one dive instructor complained to me that fishers constantly stole the marker buoys and the ropes for the wrecks, and he was the one who had to constantly replace them. He doubted whether residents had the energy or capacity to follow up on this. Other dive instructors voiced their frustration when they saw local fishers fishing in the MPAs while waiting for the dive boats to turn up. As soon as the dive boats arrived, according to these instructors, the fishers would then request user fees from the boatload of tourists. They argued that if the only motivation for having these protected areas was for money, there was little incentive to actually protect the areas, and that the conservationist 'spin' that had been placed on these projects was nothing more than a sham hiding the desire on the part of the local government to make money out of the dive industry. They pointed out that if the local government really wanted to conserve the areas, they would reduce the numbers of pearl farms, which were located right next to and in some cases within the MPAs. Although to my knowledge no scientific studies have been conducted on the polluting impacts of these pearl farms, some dive operators alleged that underwater visibility on the wrecks had decreased significantly since the pearl farms began production. For many of those in the dive industry, the issue of concern was not so much the imposition of user fees per se, but what was to be done with the money.

For other members of the dive industry, another more general argument that they voiced against the MPAs was that the ocean should be completely open access. From this perspective, nobody, not even the government, had the right to place user fees on bodies of water such as reefs because they were publicly owned. Therefore, these dive operators argued, if divers had to pay a ₱100 tax for simply diving and looking at the wrecks and reefs—taking nothing from them—fishers should have to pay a much larger tax for what one dive operator termed 'stealing fish from the ocean'.[10]

Many of these concerns of the tourism operators about the MPAs were legitimate, and clearly and honestly communicated. Similar concerns have been reported in other areas of the Philippines (Depondt and Green 2006; Oracion 2006). Their concerns however, were closely related to and tapped into more deeply held feelings about Filipinos and the Philippines, and fishers in particular, which they had built up over many years of living in the country. Such feelings varied, and became even more diverse as more and more tourism operators arrived each year. For some tourism operators, however, fishers were typically depicted as somewhat of a caricature that could be seen among many foreign expatriates, and indeed among many more well-off Filipinos. Fishers are ignorant, they are poor, they will do anything for 'easy money', and left to their own devices they will inevitably deplete the seas of every last fish. As I introduced in Chapter 2, fishers from the Visayan region, in particular, are subject to this stereotyping. Those in the tourism industry who had lived in Palawan for many years spoke pessimistically of the direction of environmental management. One particularly despairing operator stated that: 'When I first came here [Palawan], the place was beautiful. The forests were covered, the reefs were still there. Now, they have completely fucked the place up'. Another said that: 'Well when I first came here you used to be able to see the little *tambakol* [tuna] jumping out of the water just over there [points to the nearby waters metres away from his resort], then a few years later you had to go a bit further over to those islands [points to islands further away] and now it's difficult to see them in the whole area anymore'.

Tourism operators were discouraged by various practices that caused them to view fishers in this pessimistic manner. The persistent habits of using cyanide and dynamite to catch fish were two such practices which directly undermined the livelihood of the tourism industry by destroying the very products they sold—coral reefs and fish. Tourism operators often talked of the 'mentality' of the local fishers, recognising that they were very poor, but arguing that they could never see past the demands of today. I was reminded of this perception when one foreign resident asked me about the attitudes of the fishers I lived with: 'So Mike, you live with these people—are there any of them who actually think illegal fishing is a bad thing? I mean are there any of them at all who actively don't do it?' The underlying belief behind his question was that all fishers were by default involved in illegal fishing, a common assumption among some foreign

---

10   During 2006, the tax on a box of commercially sold fish was just forty pesos.

residents in Coron. Arguing that education was the key to changing long-term attitudes, one Filipino tourism worker and conservationist said that '[t]alking to these fishermen … it is like teaching a small child. At first they don't know why something is bad, but eventually they learn'. The quotes expose a conviction shared among many tourism operators that combined with poverty, the main problem in fisheries governance was ignorance and a narrow-minded hunger for resources that was believed to motivate poorer fishers. As one dive operator stated, he was very aware that he lived in a developing country and there had to be development, but he wanted the development to be 'smarter'. He asserted that if the 'mentality' of these Filipinos were to change, 'smart' development would be possible.

Those in the tourism industry thus reframed the moral appeals of those in the fishing communities as symptomatic of a 'mentality' that emphasised victimhood and reinforced a culture of poverty. Claims about a social vision of equality by Filipinos were contemptuously treated as wishful thinking, hopes that, in the absence of hard work and creative thinking, the benevolence of foreigners would support them. The foreigners saw MPAs as a typical example of the 'crab mentality', whereby the only way to get to the top is by pulling someone else down.

The local government was understood in even less favourable terms than the fishers. 'A bunch of inbred crooks' was how one tourism operator characterised the local municipal council (*sangguniang bayan*), referring to their penchant for dynastic, alliance politics which meant that most of the members were related to each other in some way. Tourism operators were even more critical of these characters, because according to them, the politicians should have been the ones to see past short-term financial goals, and try to manage the marine areas in a proper and sustainable fashion. A common criticism of the local government was that they were essentially trying to 'have it both ways'. They wanted to have the appearance of having protected areas and conservation, but they were not prepared to actually seriously protect these areas, nor were they prepared to limit any potential revenue from the allegedly polluting pearl farms: 'They want to have their cake and eat it too', as one foreign resident suggested.

Personal Investments and Territoriality of Tourism Operators

Many of the concerns among foreigners in the tourism industry about user fees, management and Filipinos can be understood more fruitfully by examining their personal contexts. When referring to the classification of those in the dive industry in his thesis, Oracion (2006: 83) asserts that 'despite their financial contributions the tourism brokers are still treated as outsiders whose attachment to the MPAs is primarily business and recreation unlike the local fishers whose ways of life or culture are directly altered by their innovations'. It is unclear from the language used whether this is Oracion's actual view, or whether it is simply the dominant local perception. Either way, I would argue that such a view is unnecessarily narrow. Such a focus on business and 'recreation' ignores the very significant personal commitments that many in the tourism industry have made in the Philippines.

For the dive operators, the oceans and especially the wrecks are viewed as 'their patch'. That is, through superior technical skills of navigation and the use of SCUBA, they have clearly marked out an area in the Calamianes where they are the acknowledged authorities. Before the introduction of the marine parks, the various dive companies had jointly managed the diving areas, setting a boat to moor every night on one of the popular reef dives, and pooling funding for things like marker buoys and ropes. They highly value the sense of local knowledge that they have about the waters of the Calamianes. Local knowledge of the wreck dives is particularly valued. Penetrating some sections of some of the deeper wrecks involves technically challenging (and potentially dangerous) diving, and instructors need a great deal of experience on these particular dives before they can safely guide other divers through them. The Calamianes has a reputation as having some of the best wreck dives in Asia, and highly experienced wreck divers from prominent diving clubs around the world will sometimes converge in the area. Being able to guide some of these serious divers through technically challenging dives is a source of pride among the more experienced operators and instructors. This sort of local knowledge can be characterised as practical, material knowledge born of experience—not the disengaged, alienated 'globe' view of the environment that Ingold (1993) has characterised as typical of the modern Westerner. As Carrier (2001) has described Westerners' personal engagements with the marine environment in Jamaica, this example can show the dangers of essentialisation and the diverse ways in which 'Westerners' can perceive their environments.

Dive operators felt that over a long period of diving in those waters day-after-day, year-after-year that they were entitled to a significant say over any rule changes to the management of the waters. This was legitimated, for example, by one of the larger dive shops naming the Esperanza reef as their 'house reef', mooring one of their boats there each night to deter fishers. As Oracion (2006: 145) points out with regard to a similar naming of reefs in Negros Oriental, calling the reefs 'house reefs' 'implies some sense of ownership by virtue of proximity and symbolizes free access' for the dive operators.

Tourism operators also fiercely criticised what they viewed as the rampant corruption in Filipino and Coron society among members of the local political elite. While they saw this evidence of corruption in many areas of life, in particular they resented the ways in which they felt they were targeted in particular, as foreign members of the society who clearly stood out. One tourist operator recounted to me his frustrations. He had lived in Coron for more than 20 years, was married to a Filipina with children and held a clear commitment to the place with his lifelong business. Despite this, he related that in public meetings and everyday life he was tarred with the allegation of being a rich, arrogant foreigner coming to Coron to exploit its resources and not leave anything for the locals. Tourism operators were convinced that, while relatives of local politicians could get away with activities such as cyanide fishing, any tiny infraction of theirs would be punished heavily. Often, they felt as if they were resources for the locals to exploit any way they could. Indeed, it is important here to recognise the complex personal histories of many of the dive operators who have married locally. In doing so, as many Filipinos put it with a big smile, they have 'married the family' as well. Most of them thus become people who are subjected to claims for resources by Filipino family members and are under intense social pressure to redistribute their relative wealth. From this perspective, the creation of MPAs can be understood in personal terms as well. The imposition of user fees tapped into a feeling among dive operators that they were being used and exploited, yet again.

For the tourism operators, the entry by the local government and communities into the realm of dive tourism management was an irritating affront to what was in their eyes, the one area of their lives where they could maintain independence, free from local corruption and patronage. Tourism more generally in Coron is sometimes represented by these foreign tourist operators as a realm of activity that can only be done by foreigners. Some talked of how local rural Filipinos did not understand the mindset of foreigners, and would never be able to provide the same level of customer service that foreigners expect: 'They need a foreigner, someone who is experienced in Western ways, to show them how to do customer service. For example, when I ask for something in a shop here, I always get the response "No we don't have". That is not OK for a Westerner! You need to apologise, make them feel better'.

These sorts of complaints were related to concerns that the system of user fees was highly uncoordinated, irritating tourists and therefore driving away the market. For example, a typical day out taking in the tourist activities could involve a trip to the wrecks, paying a ₱100 bill, then a trip to a coral reef MPA, paying another ₱100 bill, or a trip to a beach on Coron Island, paying another small bill, or a trip to the hot springs, paying yet another small bill. Their fears were that the imposition of MPAs would actually drive away tourism because of what they viewed as the inevitably disorganised and intrusive method of implementation by local Filipinos.

During the early part of 2006, tensions about the MPAs were overt. Some dive operators maintained an informal boycott on some of the reef MPAs, visiting only the wrecks, which formed the core Calamianes dive sites. 'Why do you want to help those fucking people?' called out one boycotting dive instructor to a non-boycotting instructor as he got on a boat to go to the Esperanza MPA. One group of dive operators mounted a claim that a part of the reef was actually part of the Ancestral Domain of the Tagbanua group living on nearby Coron Island, and so refused to pay user fees when visiting this part of the reef. This alerted the Tagbanuas to the possibility of claiming territoriality over the area, and they subsequently began demanding user fees from the dive operators as well. The issue was settled only when the MPA Ordinance was hastily amended later in the year to more explicitly include this section of the reef, thus ensuring that user fees for the MPA would be required. As the year progressed, the informal boycott grew less effective and it appeared that dive operators began to accept the inevitability of the situation. A lack of unified and clear representation hindered any possibilities of offering effective resistance, and while grumbling continued at the bars and dive shops, there was little they could do as long as the LGU was committed to maintaining the MPAs.

The creation of the MPAs was seen by foreign tourism operators as an entry into the one of the few spheres of life where they had managed to maintain relative freedom, control and dominance. In this way, the desire for the dive operators to maintain their management system over the MPAs can be partly seen as an assertion of both their physical, and metaphorical, territorial claims over marine territory and marine tourism and helps explain their cynicism about the potential for successful management. Claims by fishers and the support given to these claims by the municipal government were rejected as illegitimate.

Local Government Unit Perspectives and Marine Protected Area Developments

Attitudes towards MPAs among local government officials were typical of their broader attitudes about tourism in the Calamianes. As I described in Chapter 2, the local government of Coron especially is interested in developing the municipality as an ecotourism hub of the Philippines. MPAs are seen as one way to achieve this: as well as any actual conservation value that may accrue as a result of the MPA, the mere act of creating an MPA gives greater legitimacy to claims that Coron is a paradise for ecotourism. Prominent events such as then President Arroyo visiting Coron and opening one of the MPAs in 2005, the Esperanza MPA featuring on a national television program during 2006, and the US Ambassador opening the San Andres MPA in 2007 gave further authority to these claims. For the LGU then, the MPAs can be seen partly as a successful public relations exercise for Coron as a part of its aims of attracting greater levels of tourism. Not to be too cynical, there were also undoubtedly members of the local council who were concerned about fish declines and felt that MPAs were one way of addressing them.

LGU officials were not unaware of the concerns that fishers had about tourism, however. Indeed, to ignore these concerns completely would be to alienate their bases of support. During interviews, some politicians in the LGU expressed concern that tourism would not be as profitable as fishing, and would not be able to replace it entirely: 'Tourism is a good thing' said one municipal councillor. 'And here in Coron, there are plenty of opportunities to promote tourism here. But we have to be careful with tourism that we don't hurt the small fishermen. They don't speak English, they don't know how to do tourism and they have no interest in tourism'. Similarly, another municipal councillor told me that 'tourism is good, but realistically it will never replace fishing completely.'

The goals of the municipality were thus to increase tourism yet minimise the impact on the livelihoods of fishers. Lax enforcement of fishing regulations by barangay and municipal officials can be understood more easily in this context. One government official in another part of Palawan, for example, related how the explicit appeals of fishers when they are caught make punishments difficult to enforce. When a local fisher was caught using a banned compressor, he brought his entire extended family into the office of the mayor and made a great show of appealing for pity, asking the mayor how would he be able to feed this family if he was fined or were to be imprisoned. While LGU officials want the benefits of tourism, they do not want to marginalise their constituents, and are vulnerable to claims from local clients. As I discuss in Chapter 7, when I examine the traditional role of politicians in the Philippines, local leaders are expected to support residents' resource use patterns in exchange for their political support (Russell 1997: 91). Abstract rules that damage the livelihoods of people who are already poor are seen as cruel and unnecessary, whatever their alternative scientific or financial justification may be. The claims of fishers with regard to MPAs, and their discourse of the poor moral fisher, found supportive listeners in their LGU in this instance. Showing pity to fishers who break the rules is a way to gain political capital, thus reducing the incentives for strong enforcement of marine protected areas.

Plate 5-2: Typical boat used to take tourists on day trips.

Conclusion

This chapter has demonstrated how the responses of fishers to the development of MPAs are framed by a discourse that emphasises the poverty of fishers, the unfairness of regulating fishing practices, and the rights of fishers to extract user fees from the tourism industry. Fishers here appeal to the members of the local government, those in the tourism industry and the NGO workers assisting with the implementation of the MPAs. If MPAs are to be instituted, they argue, they should have minimal impact on existing fishing practices, and should actively provide benefits to fishers. MPAs are viewed here as interventions that should assist the livelihood of poor fishers.

The ways in which MPAs in the Calamianes have been contested has some important policy implications. MPAs are often viewed as an excellent tool by conservationists and, less often, by those coming from a perspective focused more on fisheries management. This is because MPAs have, in theory, both the strict protection zone so favoured by conservationists who wish to protect biodiversity, and also the spillover effect, which appeals to those who wish to increase the level of fish stocks outside of the MPA. They therefore aim to satisfy all stakeholders, including the tourism industry and fishers. In this chapter, I have shown that in the Calamianes during 2006–07 at least, such a characterisation is not correct.

The aims and motivations of fishers, dive operators and conservationists are very different. As Oracion (2003) reported for a similar situation in Batangas, the original motivations for supporting an MPA among local fishers were overwhelmingly (95 per cent) because of the potential they saw for a 'sustainable subsistence fishery', as opposed to resource conservation, which drew zero responses (ibid.: 113). The responses of regulators and private tourism brokers however, focused on the potential for resource conservation and tourism revenue. Similarly, Christie et al. (2003: 319) have argued that the motives of different stakeholders are closely linked to the sustainability of coastal resource management institutions. They point out that where interests and motives clash, the potential for conflict may exist after the project or organisation implementing the MPA has left the site.

In the MPA meetings and discussions, I observed in the Calamianes, the aims of fishers with regard to the MPA were strongly directed towards what they saw as an opportunity to protect their patterns of exploitation. This is primarily because of their representations about fishing—legal fishing is harmless, moral and associated with poverty; illegal fishing is harmful to the environment, immoral and associated with a higher level of wealth. Fishers claim that their activities have a negligible effect on the environment and that MPAs should be about taxing those resource users rich enough to afford it (the tourists), basing their arguments on a vision of social fairness. These perceptions have influenced the process of implementing MPAs in several ways. Core zones have often been minimised as much as possible and seen as a concession to conservationists, buffer zones have been adapted to include the fishing techniques and gears of local fishers, and enforcement has rarely been effective when it has been conducted by locals.

In contrast to the fishers, dive operators have understood the role of MPAs as primarily about conservation. From this perspective, any revenue gained from user fees should go towards management and enforcement costs. Strongly affected by the transfer of management to communities, and the imposition of user fees, their responses were very negative. They argued that the communities' approach to MPAs was not about conservation at all; this was just a sham hiding the real desire among local government officials and communities to make money out of the dive industry. The dive operators framed their argument about the impacts of their actions in direct opposition to the arguments of fishers. They argued that in contrast to the fishers, who 'steal fish from the ocean' and indulged in irresponsible environmental behaviour, diving activities had no negative environmental impact and so should be free of charge. They based their arguments on the logic of good environmental management, and I have demonstrated some of the underlying territorial concerns that also lay behind these arguments. For the LGU, MPAs have been seen as a way in which to expand its own territorial claims over tourism, and so MPAs have been generally enthusiastically supported. Within this supportive context for MPAs however, fishers' claims have also been privileged, thus reducing the capacity for effective enforcement of the MPAs.

The status of MPAs in the Calamianes could be seen during 2006–07 as somewhat ambiguous—many were being created, but all of them were marked by serious limitations. These shortcomings and ambiguities were a direct result of the discourse of the poor moral fisher that formed the basic perspective of fishers when it came to negotiating MPAs. Whether the MPAs are able to produce the kind of increase in fish stocks desired by the conservationists remains unclear because of these shortcomings.

From this perspective, the proliferation of MPAs can be understood not so much as a victory for conservation or wise fisheries management, but as a way in which fishers and local government officials have been able to successfully assert their territorial claims and advance their interests. The development of MPAs in the Calamianes is fundamentally a social process, one that is shaped by specific ideas about poverty and the environment, and the local political contexts in which these ideas are expressed.

6. Fishing in Marine Protected Areas: Resistance, Youth and Masculinity

In the Calamianes, certain fishers continued to fish within MPAs, and expressed the view that they had a moral right to do so. These fishers represented fishing within MPAs as a form of resistance against unjust regulations, a topic related to the themes of the poor moral fisher. While these fishers did not always present themselves as pitiful, their justification of this form of fishing presented a 'basic rights discourse' that emphasised the value of fairness and the right of fishers to fish in all locations. I also argue, however, that only certain fishers deliberately worked inside the MPAs—younger men, whose ideas about fishing involves notions of masculinity and particular economic and personal values. This chapter will analyse fishing within MPAs by examining two related motives: resistance against regulations, and the desire to demonstrate a vision of masculinity that is intertwined with specific notions of personal and economic success.

As I indicated earlier in this book, the term 'illegal fishing' covers a wide range of practices. While the type of fishing I focus on in this chapter—fishing within MPAs—is technically illegal, it is not understood as illegal in the same way as cyanide and dynamite fishing. This is an important point because I wish to show in this chapter how the discourse of resistance among those who fished within MPAs emphasises the immorality of cyanide (and dynamite) fishing, as does the discourse of the poor moral fisher. However, the second motivation of these MPA fishers has links with cyanide fishing. Both forms of fishing emphasise high risk and high return, masculinity and the desire for economic success. I will show how fishing within MPAs can be interpreted as a form of fishing that lies between the categories 'legal' and 'illegal'.

The chapter first reviews some of the literature relating to various forms of illegal fishing. Here, I contrast some of the material that has concentrated on understanding why illegal fishing flourishes in the Philippines, with literature from both the Philippines and elsewhere that addresses motivations behind illegal fishing and 'poaching'. I then detail the discussions I had with those who practiced fishing within MPAs and tease out the links between fishing and masculinity. In the main body of the chapter, I focus on how younger fishers are more likely to take up illegal fishing because of their particular economic and personal values. Younger fishers, I argue, are more interested in higher-risk fishing and ultimately in moving away from fishing itself.

Academic Approaches to Illegal Fishing

Various analysts have looked at why various forms of illegal fishing have continued to flourish in the Philippines. In San Vicente (Palawan), Eder (2003: 214) points out that beach seining, an illegal gear, 'enjoys considerable local tolerance because even those who do not engage in beach seining themselves may benefit from its presence'. The fish that are caught in beach seining, for example, provide those raising hogs with a useful feed source in the form of fishmeal. Similarly, Russell and Alexander (2000: 33–4) relate how share systems among fishers serve to provide community support for blast fishing in the Lingayen Gulf. There, fishers using dynamite regularly give away portions of their catch to small-scale gillnet fishers. Galvez et al. (1989: 49–50) also demonstrate how various members of the village they studied (also in the Lingayen Gulf area) benefit through blast fishing: either through seasonal employment, or fish giveaways. They also stress the role of local law enforcement agents in exacting bribes, and that of local politicians in suspending specific cases of law enforcement in exchange for political support.

Other writers have looked at the motivations behind illegal fishing in the Philippines. That it clearly appeals as a rapid source of income is obvious: years ago Spoehr (1980: 24) pointed out that '[d]ynamite fishing was the quickest and most economical way for small-scale fishermen as well as middle level operators to increase their catches, and fishing communities rapidly adopted dynamiting as a general practice'. Similarly, Szanton (1971: 30) noted that in the northeastern corner of Panay where he worked during the 1960s, the 'potential earnings ratio [of dynamite fishing] is too high to be ignored by some of the local fishermen even if it means the ultimate destruction of their fishing grounds'.

In addition to the obvious financial motives, however, other writers have suggested that many illegal activities are a form of resistance against unjust regulations (Scott 1985). Lahiri-Dutt (2003) provides an example of such a perspective in the context of illegal small-scale 'informal' coal mining in Eastern India.[1] She asks, '[w]hat factors force ordinary humans to turn into unlawful citizens or criminals' (ibid.: 69), and argues that resistance was the key to understanding such activities:

> In my view, this practice is one way of re-establishing the lost claims of the local communities over the land and its resources. When the local communities found themselves disempowered by the laws, and

---

1   Although Lahiri-Dutt is referring to claims over land, MPAs zone particular spaces according to what sort of fishing is or is not possible and so the comparison is still highly relevant.

disenfranchised by the mining company, they began to extract as much as they could from the same land, the land they can no longer call their own (ibid.: 75).

Similarly in Greece, Bell et al. (2007) observed poaching in a lake and understood it to be a response to a series of environmental regulations that were deemed unfair, hypocritical and benefiting farmers who lived far away while marginalising the livelihoods of local fishers. 'In this context, poaching becomes emblematic of lost livelihoods and identities and a token of resistance and rebellion' (ibid.: 401).

Importantly for the purposes of my argument with regard to the different ways in which fishing with cyanide and MPA fishing were represented locally, Bell et al. (2007: 399) also argue that '[s]ome people who admit undertaking what they perceive as least detrimental (sic) forms of poaching are antagonistic towards what they construe to be truly harmful forms'. Similarly, for fishing in Lithuania, Hampshire et al. (2004: 313) note that '[p]oachers occupy different positions along the spectrum, according to their insider/outsider status, their perceived need or greed, the apparent threat posed to fish stocks, and the aesthetics and fairness of their fishing practice'. In a situation similar to what I observed in the Calamianes, they adopt a perspective where local subsistence anglers are contrasted with non-local electro fishers.[2] In the Philippines, Russell and Alexander (2000) and Galvez et al. (1989) both report that fishers in the Lingayen Gulf emphasised that blast fishing was relatively harmless compared with illegal commercial trawl fishing, thus justifying their actions. Eder (2008: 112–3) also describes how beach seining is not considered as illegal as cyanide and dynamite fishing in San Vicente.

My understanding of fishing within MPAs resonates with the insights of these writers, who emphasise how these forms of fishing can be understood as resistance against unfair regulations, and are contrasted with other more damaging gears. In a context where restricting the activities of legal fishers is seen as unjust and exploitative (as I have argued in Chapters 4 and 5), fishing within MPAs is one defiant response. Some fishers felt justified in fishing within MPAs, arguing that their gears should not be subject to any restriction (as opposed to the use of cyanide and dynamite).

However, from my limited discussions on the topic to follow, I formed the view that it was definitely only a small minority of fishers who actually fished in these MPAs. Therefore, I also want to examine how and why younger fishers in particular adopted this practice. As I shall show, notions of bravery, status and the pursuit of wealth were also important for those young men. I argue that in

---

2 Electrofishing uses electricity to temporarily stun fish.

addition to notions of resistance, fishing within MPAs can also be interpreted in the context of a web of local concerns that encourage young men to demonstrate their masculinity.

On this second point, my understanding corresponds with what Lowe (2000: 246) describes among young cyanide fishers in Indonesia:

> Nearly all cyanide practice is carried out by young men, for several reasons. Older people have trouble diving the way younger folks can; it is a physically strenuous activity and older men complain of the cold. High live-fish profits are a way for young men to build houses and establish new, independent families. Cyanide also has a status that is appealing to younger people. Cyanide fishers peacock their wealth and modernity by controlling outboard motors. They also have the money to smoke expensive cigarettes and wear new clothes. That the activity is illegal further demands their daring and indicates their tightness with officials who will protect them from prosecution. In short, cyanide fishing is where it's at—what's happening—and this makes it a game young guys want to play.

Fishing within MPAs is an interesting mode of fishing because while those who practiced it contrasted it strongly with cyanide fishing, the visions of masculinity that were articulated are strikingly similar. I suggest therefore that fishing within MPAs occupies a 'middle ground'[3] in the spectrum between completely legal fishing and the 'truly' illegal fishing such as cyanide and dynamite. It has characteristics that link it with both, so in this chapter I seek to analyse how fishing within MPAs is best understood from perspectives of both resistance and masculinity.

Fishers in Marine Protected Areas

Resistance

As I introduced in the previous chapter, a majority of the MPAs in the Calamianes are associated with tourism. When discussing fishing within MPAs, fishers emphasised the MPAs that surrounded resorts. When I asked those who admitted to fishing there how law enforcement officials and tourism operators might feel about such activities, they were highly critical of the situation.

---

3   See Turton (1986) for a similar discussion of the 'middle ground' with regard to everyday peasant resistance.

Operators managing resorts in an MPA were usually described as rich foreign businessmen (or rich businessmen coming from Manila) who made a great deal of money out of the beautiful reefs of the Calamianes. They felt that these businessmen were given preferential treatment by the local government because of their money. Fishers resented the fact that they were denied access to the reefs, which in some cases the resort operators had cordoned off as MPAs. As I argued in the previous chapter, the potential benefits of MPAs for fisheries are not widely understood or accepted among local fishermen. Instead, MPAs were often viewed as further marginalising the meagre livelihood of small fishers. The emphasis devoted to tourism by the local government was seen by many, as a trend that would encourage the development of more MPAs, which would make it harder for them to make a living through fishing. In their view tourists, businessmen, and well-connected officials in the local government were frequently the only people who benefited from the declaration of such protected areas. They felt, therefore, that they were justified in continuing to fish there.

While I did not take notes at these discussions, I recorded my impressions shortly afterwards and comments such as the following give a flavour of the tone:

> At Resort X, they banned fishing around the whole island. This is completely unfair.

> They have banned some of the best fishing grounds.... I used to fish at Donido, that was the best spot to find grouper; but the resort has made it stricter now.

> The government is making it harder for us poor fishers but they always benefit the rich. Why should the rich tourism operators be allowed to do this?

Law enforcement officials were routinely described as blatantly corrupt, parasitically seeking to take advantage of poor fishers (see Chapter 7). They were depicted as facilitating cyanide fishing through bribery, and oppressing poorer fishers who practiced legal methods. As I described in Chapter 4, among many fishers there is a broad feeling that they are the poorest of the poor, and have been abandoned by the government. In this way, the activities of people trying to stop fishing within MPAs are delegitimised as corrupt and exploitative.

Enacting Masculinity

The context in which discussion of fishing within MPAs was brought up epitomises the ways in which some young men in coastal communities perceived it. From my experience it was always during a *tagay* session, when young men sat around in a circle and took shots from whatever alcohol was on hand, usually a bottle of rum (Tanduay) or brandy (Emperador), or less frequently beer. *Tagay* is a social event that not only reaffirms togetherness (*pakikisama*) among one's friends or peer group (*barkada*), but also reaffirms a sense of manhood or machismo among those who can drink heavily and hold their alcohol. The *tagay* sessions that I was involved in were usually composed of men only, and the dynamics were usually very jovial and full of laughter, combined with a pervading sense of machismo. The fact that this form of fishing was discussed only within an atmosphere of machismo, such as *tagay*, suggested a link to be explored.

When fishers did talk about their experiences with fishing inside MPAs, it was referred to very much as an example of strength or bravery. For example, one young man talked with pride of how he was able to avoid the guards deployed at a private, foreign-owned resort island, and fish within the surrounding MPA. Others spoke of their skill in avoiding the state employed guards in other MPAs. All the fishers I spoke to who talked of such fishing declared his practices in terms of a bold, brave act in which he was able to avoid capture; nobody was able to stop him from obtaining a large catch of fish and the ensuing financial rewards. In this sense, the act of fishing within MPAs can be seen as an active 'gamble' with the state (Aguilar 1998: 32–62) and outsiders that usually tended to pay off. Importantly, every single person I talked to who was engaged in fishing within MPAs was a young male in his teenage years or in his early twenties. I turn now to a discussion of how fishing generally is intimately linked to notions of masculinity in Esperanza.

Links between Fishing and Masculinity

In Esperanza, fishing can be understood as both a livelihood and a practice that is connected to various ideas surrounding notions of masculinity. Although fishing itself is not an exclusively male affair; men dominated it. Women do not participate in commercial fishing trips, and only rarely are they seen on shorter net-fishing trips. Siar (2003) shows how in Honda Bay, on mainland Palawan, women's fishing activities are closely associated with the intertidal zone. In Esperanza, similarly, women contribute significantly in practices such as gleaning for shells, catching shrimp, and post-harvest selling of fish and transportation. However, women will often downplay the worth of these contributions, and indeed are often embarrassed about them.[4] So, even if in reality the work of fishing may be shared between the sexes (Weeratunge 2010), the point remains that fishing is associated with an ideology of masculinity.

Like other all-male activities, such as attending a cockfight (Aguilar 1998: 32–62), fishing is a gamble and an opportunity for male fishermen to demonstrate their masculinity, economic prowess, and value.[5] Commercial fishing trips in the region last for up to two weeks; during this time up to 30 men will eat, fish, and sleep on the boat together. On their return, the drinks are invariably broken out as soon as the packing and transportation process is completed. As discussed earlier, this is an opportunity to reinforce the comradely male bonds that exist during fishing trips, in an atmosphere pervasive with machismo. Those men who can pull in the biggest or the most fish during these trips are highly respected among their peers. The fishermen who had managed to achieve relative success in their profession and own several commercial boats are the most respected men in Esperanza, the most financially well-off, and hold most positions of political authority. Although in broader Philippine society fishing remains a low-prestige occupation, it is one that can still bring relatively high status within a fishing community (Russell 1997).

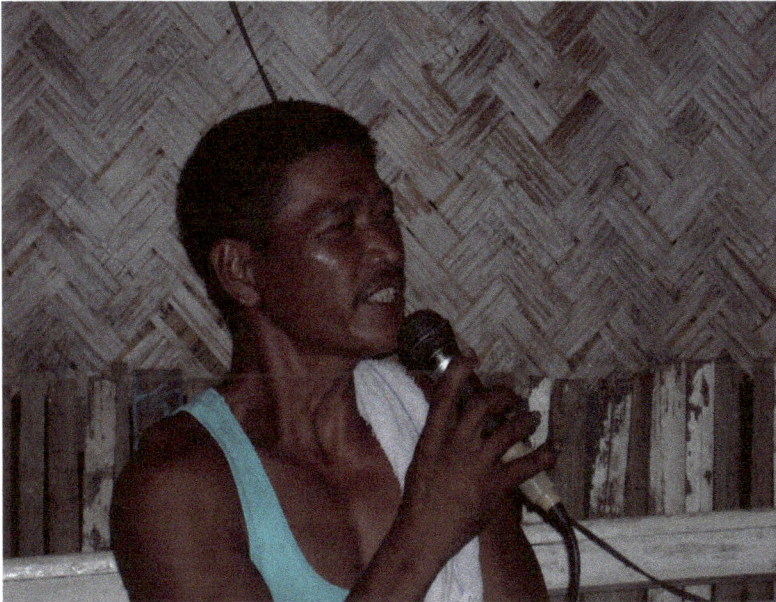

Plate 6-1: At the karaoke den.

4   Swift (2006: 6–7) suggests that in the case of the seafaring industry in the Philippines, a similar practice of downplaying the worth of women's economic contributions is related to the potential for capitalistic exploitation of female labour.

5   Indeed, along with the karaoke den (Plate 6-1), gambling is one of the favoured recreational activities among many fishers when they return to shore. Fishers will not necessarily wager large amounts of money, but will spend a lot of time playing cards or attending cockfights.

What I observed in Esperanza is similar to what Russell (1997: 85) has described in a Batangas fishing community, where:

> Being captain of a boat provides a man with rich opportunities to demonstrate masculine skills, bravery and ability to attract or manipulate mystical sources of luck and potency…. Boat ownership as a form of symbolic capital transcends its utility as a source of production, since it also expresses a distinctly masculine identity in terms of one's ability to physically withstand the rigors of a sea-going occupation and lifestyle.

The example of Manuel gives some idea of how fishing is viewed as a vehicle for masculine expression in Esperanza. When talking of his experience as a captain of a fusilier-fishing boat, he described his skills with great animation, hand motions and rhetorical flair:

> I am not boasting or being arrogant, but I am the best fusilier-fishing captain in the community. It is true. You ask your kuya [older brother] Bong, he will tell you the same. My catch is always 2200 kg, 2400 kg. Last year my lowest catch for the whole season was 1500 kg! Most of the other captains, they will only catch 1600 kg, 1700 kg, 1800 kg, like that. You see, when I am out there, I don't waste a single minute. During the day, as soon as the sun is up I make sure everyone is busy fishing. As long as the weather allows, everyone must fish. I don't stop for food breaks—my crew eats while they fish. During night, ok then they can sleep. But I am the captain and I don't allow anyone else to pilot the ship during the night, so I travel at night to go to the good fishing grounds. Other captains just anchor at night and then move around to the fishing grounds during the day, so they lose lots of fishing time. When I drive the boat I have a lot of coffee, I turn the radio up loud, I smoke lots of cigarettes to keep me going. Sometimes when I get back to Esperanza people are surprised because I am back so quickly, but I've already caught enough fish.

Tall, extremely strong and exuding self-confidence, Manuel was one of the most highly respected fishing captains (in any fishery) in Esperanza. He complemented his fishing exploits with considerable skill as a carpenter, and as a soulful crooner of epic ballads in the *videoke* den. Other fishers in Esperanza would speak with admiration of the skill and experience of captains like Manuel and Carlos Marquez, the fresh grouper fishery captain whom I have described in earlier chapters:

Kuya Carlos, we call him the 'Professor of the Sea' here because he has taught so many of us how to fish. Many of the techniques that we use here in Esperanza were first brought in by Kuya Carlos from his relatives in Mindoro…. Kuya Manuel is really good with fusilier fishing because he knows the currents and the fish better than anyone else. He can predict where the fish will go.

Dumont (1992) describes a similar understanding of fishing in the village on Siquijor where he worked during the 1980s. Stating that '[t]he ability to catch fish was an expression and a measure of male success' (ibid.: 112), Dumont emphasises the ways that men use their fishing success as a way of gaining prestige. Fishermen give away fish in order to support an exchange network, and to be generous is considered 'compulsory behavior for any fisherman' (ibid.: 115). Fish is not only good food to eat but

also food with which to play, to display, and to act out a fisherman's aggressive and competitive *buut*, his "identity". For indeed, sending fish here and there constantly to everyone and anyone, generous as it may have been, was in addition, in supplement, a subtle way of showing off, of bragging about the results of one's efforts' (ibid.: 115–6).

Dumont went further to suggest that fish was the icon of phallic aggressiveness that 'pointed up the keen competition that existed among the fishermen' (ibid.: 116).

This description of the ways in which fish are shared represents similar patterns of social reciprocity in Esperanza. When net fishers arrive in the mornings, the beach is very active as neighbours come by to inspect the catch. Fishermen who bring in a good catch hand out fish to those who pass by: *ulam mo* (your dish to go with rice) they would say, pointing out particularly fine specimens. Other people, such as elderly men no longer able to fish regularly, frequently help with the processing of the catch and earn themselves a few generously sized pieces of fish for their help. When the time comes for *tagay*, men will be pleased if they get a chance to provide the food that goes with the drinks (*pulutan*). This was usually grilled fish (see Plate 6-2). In this fashion, sharing of fish can be viewed as one of the ways that masculinity is enacted in Esperanza.

Plate 6-2: Freshly caught fish to go with drinks.

From such accounts fishing is clearly a practice intimately tied to understandings of prestige and masculinity. However, there is an important variable within the category of 'males' that is not always considered when the links between fishing and men are considered—age. The remainder of this chapter argues that age has considerable implications for how fishing within MPAs has been practiced.

Younger and Older Fishers

Hook-and-Line Fishing Versus Net Fishing

In Esperanza there is a marked difference between younger and older men in terms of how they conceive of and approach fishing. This difference can be seen through a brief examination of the different types of fishery that each group favour. Most of the older men in the community (those aged above 40) prefer working on small-scale net-fishing boats. These men have invested considerable money and effort over many years in getting to where they are, with their own motorised boat and gillnet, which they use to catch rabbitfish. They catch these fish in the seagrasses close to shore. As I described in Chapter 3, trips usually only last one day, and the fish are usually sold at the market in Coron town.

In contrast, younger men can be seen commonly on two types of fishery—the hook-and-line fresh grouper fishery, and the live fish or live grouper fishery. The captain and a few select others in positions of relative authority, such as the engine-operator and the ice-hold-packer (Plate 6-3), are usually older and highly experienced in the commercial fishery of fresh grouper; however, the crews of both grouper fisheries are dominated by teenagers and unmarried men in their twenties. Although it is usually only the live grouper fishery that is implicated in fishing within MPAs, such crews are often interchangeable so that a young man may be a crew member of a fresh grouper boat one season, but join the crew of a live grouper boat the next. The point is that it is mostly young men on those hook-and-line boats, and those young men have specifically different ideas about fishing and the financial rewards available to them compared with older men. Such ideas are embodied within the different styles of fishing each group chooses to adopt.

Plate 6-3: The ice-hold packer aboard the fishing boat.

One important difference between the hook-and-line fisheries and net fishers is the level of stress or challenge involved. The hook-and-line grouper fisheries have extended trips of between ten and seventeen days, which can be quite physically and mentally challenging. For example, fishing for fresh grouper involves leaving the mother boat on a small one-person boat, which can be an extremely isolating experience if the fish are a long way out in open seas.

Fishing for live grouper includes the responsibility of looking after the fish after they were caught, and effectively 'babysitting' them in the aquariums built into the boat.[6] Crew members cited fundamental hardships, such as being away from the family, lack of sleep, the basic living conditions on the boats, and the danger of heavy seas during rough weather.

Net fishing, in contrast, is considerably more sedate than the hook-and-line fisheries. In response to a question on why he preferred net fishing to fishing on the large commercial boats or on live fish boats, Bong, a fisherman in his early-fifties, pointed out that net fishing was altogether safer and less stressful for him than the commercial grouper fishery or the live reef fishery. It was closer to shore, it was physically less demanding, and the trips were shorter. Bong explained that metaphorically speaking, his life was in the 'late afternoon' now (*hapon na*) and his focus was on spending and enjoying time with his family; he would leave the hard methods of fishing to the younger men, whose lives were still in the 'morning' (*umaga pa*).

Another important difference between net fishing and hook-and-line fishing is the pattern of earnings. Overall, net fishing is regarded as a fairly consistent and reliable method of catching fish, where even a poor catch can still bring in rabbitfish worth ₱100 or so. In contrast, the method of hook-and-line fishing is much more variable, and catches tend to fluctuate erratically. While rabbitfish are easily collected en masse in identifiable habitats that are close to shore (that is the seagrass beds), groupers are brought in one at a time from wide fishing grounds. The fish targeted by the hook-and-line fishermen are known locally as 'first-class' fish: groupers, live or dead, are the primary target, but there is often a by-catch composed of other large, well-priced fish such as tuna and snapper. All of these fish tend to be large and impressive looking compared with the smaller, brown, and spiny rabbitfish (Plates 6-4, 6-5, 6-6). The price of live grouper varied markedly each week, but often reached around ₱2000 per piece when sold to the traders in Coron town in 2006–07. The fresh grouper caught on the commercial boats sold for around ₱600 per kg in 2006–07 in Manila, which was significantly higher than the normal price of rabbitfish (around ₱40 per kg in Coron) obtained by net fishers. The potential for high earnings, therefore, is greater in the hook-and-line fisheries, but the income from net fishing is altogether more stable.[7]

---

6   This responsibility was explicitly cited by one older fisherman as to why he did not enjoy live grouper fishing: 'With this sort of fishing you have to babysit the fish! You can't sleep properly; you always have to be checking to make sure the fish are OK; if one gets sick you have to watch it quickly to make sure the sickness doesn't spread to the other fish; it is very hard work! I much prefer just to catch the fish and throw it in the hold'.

7   Such a contrast corresponds with what Eder (2008: 75) observed for one fisher in San Vicente. For this fisher, the returns for squid were varied and unpredictable, but sometimes 'hit the jackpot'. Bream, however, pulled in a lower but steadier income. Eder points out that this fisher had to choose between squid and bream

Plate 6-4: Fisher with tuna (Family Scombridae).

fishing whenever he decided to fish, and that '[i]n making this choice, his main consideration was his ability to tolerate the risk of catching nothing at all' (ibid.). I would suggest that such a consideration would be influenced heavily by the economic and social values of the individual, as I argue in this chapter.

Plate 6-5: Rabbitfish (Family Siganidae).

Plate 6-6: Fresh grouper (Family Serranidae).

Fishers cite these different patterns of earnings as good reasons to fish the way they do. Older net fishers, who have to support a family by bringing in a relatively reliable source of income, state that they prefer the stability and consistency of net fishing. They recognise that hook-and-line fishing can be more lucrative than net fishing, but they stress that, on a bad day, those engaged in the former can come back with very little or no catch at all. 'They only have jackpots' (*Jackpot-jackpot lang sila*), some net fishers say, meaning that the hook-and-line fishers simply obtain bonanzas and have no security. They also note critically that most hook-and-line fishers have to travel further and for a longer time. Thus, expenses are much higher than in net fishing, and so the need to obtain a good catch and the potential for a big loss are greatly heightened. Additionally, they point out that the credit arrangements for the hook-and-line fisheries tend to be onerous (see Chapter 3). Whereas most of the net fishers in Esperanza either own their own boats, or work on boats owned by a close friend or relative, most of the hook-and-line fishers work on boats obtained through a loan. The latter are obligated to supply all their fish to the trader who has loaned the boat to them, and to pay off gradually the debt for the boat and engine. Because of the financial hazards of hook-and-line fishing, therefore, the older men in the community prefer net fishing, which yields a steady and assured income.

Younger fishers, however, point to the potential of getting the 'windfall' or 'jackpot' through hook-and-line, especially the live grouper fishery: 'Even if you only catch a few pieces, this is still enough to make a profit if they are kept in good condition. The price is much bigger'. The hook-and-line grouper fisheries can be seen as a far riskier bet than net fishing, but one where the payoffs are potentially much higher.

Economic and Personal Values

The different motivations that surrounded earnings from fishing are pointers toward more general understandings of the ways that younger and older men conceptualise the relationship between their livelihood, earnings and personal lives. Many of the older fishermen I interviewed said that their social position as family men had been settled for a long time. Bong, for example, said that while he recognised that he was quite poor relative to many people such as those living in Coron town; he had accepted this situation. The values he held dear and emphasised to me were those of family. Many other older men in Esperanza similarly expressed to me that, although they were poor, they were happy and contented with their families. One episode during my fieldwork illustrates perfectly the differences between older and younger men in regard to attitudes toward money. An older fisherman (a net fisher) told me that he earned about ₱4000 (US$80) per month, which he said was not much, but was enough for him to raise a family and buy food to eat. In response to the same question on monthly income, his son (a hook-and-line fisher) also quoted a figure similar to that of his father. In contrast to his father, however, he went on to explain how inadequate his income was for the goals he wanted to pursue, such as going to college and finding a good job in Manila. While the older man identified with the notion of being a 'fisherman', his son viewed fishing more as a (hopefully temporary) means or standby job to obtain a limited income than as a life calling.

The ways in which a bachelor lifestyle tends to influence the lives of many young men is important in Esperanza. Here, my analysis bears a resemblance to the descriptions of Lewis (1992: 202–4), in his ethnography of vegetable farmers in northern Luzon. Lewis describes a class of young men in terms of a 'youth culture' that developed in the wake of the decline of the vegetable industry and the growing scarcity of land. Lewis emphasises the ways in which a boy's peer group or barkada tends to chart the pattern of his behaviour, so that they begin to become part of the bachelor lifestyle while still in elementary school.

Lewis's emphasis on the bellicosity of youths in northern Luzon is much stronger than what I saw in Esperanza, where many of the young men continue to live, on the surface at least, nominally respectful lives with their families. However, some of his observations about the ways in which the bachelor lifestyle tends to structure patterns of behaviour certainly hold true for what I observed in Esperanza. For these young men, many of whom have not had the opportunity to get a high level of education, attaining a certain level of status within their peer group is very important. The young men in Esperanza stick together closely, and participate with great enthusiasm in activities like basketball, drinking sessions, and hunting for 'chicks'. Material objects of status are highly valued in this context, and so income becomes a means to enhance their status within this peer group. Money thus tends to be spent on flashy consumer goods such as motorbikes, alcohol and clothes. This corresponds with the observations made by Galvez et al. (1989: 50), who reported that:

> [t]he very little surplus many fishermen get [through blast fishing] is spent in other "get-rich-quick" schemes similar to blast fishing, such as *hueteng* (an illegal form of small-town lottery), *ending* (a form of lottery where the winning combination depends on the results of the national basketball games), and other kinds of gambling activities.

The ability to obtain a relatively high level of income is also of course important for its demonstration effect outside the peer group—especially to potential future life partners.

As is argued elsewhere (Bulloch and Fabinyi 2009), the idea of an attractive male partner for marriage in the rural Philippines tends to be based around a range of preconceptions and ideals. An unofficial continuum of beauty widely exists among residents in Esperanza and elsewhere in the rural Philippines that strongly correlates with class and race. Pale skin, a prominent nose, straight hair and tall stature are considered attractive, whereas darker skin, a flat nose, curly hair and short stature are held to be unattractive. Therefore, at the bottom of the hierarchy of physical attractiveness are members of indigenous minorities, particularly those groups with especially darker skin and curly hair. Then follow working class Filipinos—fishers and farmers—whose skin is browned from work in the sun. Then, well-off Filipinos with paler skin, Mestizo Filipinos, and Caucasians.

Financial security is also extremely important in marriage considerations, as indicated by the common saying in the Philippines 'no money, no honey', or as one younger male fisher in Esperanza put it: 'Here in the Philippines it doesn't matter if you are ugly. If you have money, many girls will follow you'. It is widely perceived that these physical and economic characteristics of a desirable marriage partner cannot be easily accommodated through a life of fishing.

As well as signifying local status, therefore, high income levels are a way by which young men in these communities can actually move outside the local and access a broader dream of success, such as that described by the young hook-and-line fisher earlier—going to college and getting a nice job in Manila or abroad. Thus, for these young fishers, the practice of fishing within MPAs also offers the potential of moving into a new, empowered space. The promise of high rewards in fishing within MPAs complements perfectly the desire for material status possessed by young men, and at the same time enhances the fishers' reputation for strength and bravery within the peer group.

From this perspective, fishing within MPAs can be conceived of potentially as a 'rite of passage' for some young men in this area, similar to that described by McCoy (1999) in his study of the Philippine Military Academy, where graduates formed close male bonds through the process of military socialisation.[8] Fishing within MPAs thus appears as a form of group socialisation that celebrates masculine values of courage, independence and bravery. Just as Dumont (1993: 423) states that 'fishing was the defining feature of an all-male activity', I argue that fishing within MPAs is an intensified expression of masculinity, with its overtones of high risks and high returns. It would be useful to conduct further detailed research to determine to what extent the internal group dynamics of these fishing

---

8   Other writers have observed similar practices throughout the Asia-Pacific region. Stacey (2007: 42) writes that '[s]ailing is almost a rite of passage for many young males' from Rote in Indonesia. Potter (1997: 301) also describes young Iban males going into the forest to collect natural rubber, 'testing one's strength against the perils of the wilderness'; Vayda (1969: 211) notes how young Iban men made various types of extensive journeys for the 'incentives of material profit and social prestige'.

activities might correlate with other accounts of masculinity in the Philippines, such as Jocano's (1975) description of gangs in Manila slum communities, or the *barkada* of Siquijor portrayed by Dumont (1993).

Links to Cyanide Fishing

The links with masculinity I have outlined above bear a significant resemblance to the ways in which cyanide fishing was discussed by those who admitted to using it. As with those who talked of fishing within MPAs, the few discussions I had about the use of cyanide only tended to arise during tagay sessions among young men.[9] While these fishers did not emphasise the values of resistance against the state or against tourism, their language emphasised skill, bravery and strength (*malakas*). For example, some spoke of bribing the Coastguard, or of their skill in evading them completely: 'We go at night time, when the moon is new, so nobody can see us. There is nothing the Coast Guard can do against us!'. Similarly, Galvez et al. (1989: 50) argue that '[n]onblast fishers fear the threat of being arrested more than the possibility of accidents. To a certain degree, we can say that those who engage in blast fishing have bolder personalities'. As with those who fished in MPAs, every fisher I spoke to who admitted using cyanide was a young male. Other residents in the Calamianes I have spoken to, including NGO workers and other fishers, supported this observation. 'They do it because they have a brave heart', as one NGO worker described it.

Thus, despite the anger against cyanide expressed among those who fished within MPAs, in its enactment of a specific vision of masculinity there is in fact a significant resemblance. Because of the limited and sensitive nature of the data, it is impossible to speculate about how many people may have actually been involved in both forms of fishing. Here I merely trace one link between the two.

Modernity, Globalisation and Alternative Futures

I have shown how the attraction of younger residents to fishing within MPAs is related to particular ideas about social and economic empowerment: moving out of poverty and out of fishing. These perceptions resonate with other practices in Esperanza, and can be conceived of as various ways in which residents have attempted to engage more fruitfully with modernity and globalisation.[10]

---

9   It should be emphasised that none of the fishers I spoke to who admitted to using cyanide came from Esperanza; they were groups of young men in Coron town.

10   In an argument that parallels my own experiences with cyanide fishers in particular, Macintyre (2008: 180) has argued that among young men in Papua New Guinea, 'aggressive masculine behaviour is implicitly valued as both an expression of engagement with modernity and as an ideal of charismatic self-assertion that is transgressive, audacious and risky'.

In a broad-ranging review of the relationship between rural livelihoods and poverty, Rigg (2006: 189) has argued that in rural Southeast Asia, there has been a significant shift in attitudes towards traditional forms of rural employment such as farming:

> Rural existences are becoming almost as monetized in countries like Thailand, Malaysia, and Indonesia as are urban lives. Farming has become, often in little more than a decade, a low status occupation to be avoided. This view has a marked generational component: it is younger people who most urgently and fervently wish to build futures that avoid farming.

Rigg cites numerous studies from both mainland and island Southeast Asia that demonstrate the prevalence of such a view, arguing that it is due to multiple factors such as education, the media and consumerism. Kelly (2000: 103), for example, reports that in Cavite in Luzon, 'young people have shifted their aspirations and expectations away from a rural life and towards other forms of work'.

In Esperanza, such a view is consistent with the preferences of younger fishers to avoid long-term investment in fishing and, ultimately, to move out of the fishing sector entirely. As indicated by the response of the young fisher I documented earlier in this chapter, many young people are keen to migrate to Manila, in order to get a good education and, consequently, get a good job. Some young people, supported by their parents, had been able to do this. The Calvino family, whom I described in Chapter 4 as not being able to persuade their older son to move out of fishing, had been able to send their younger son to Manila to enrol in a mechanical course. From this, they hoped he would be able to find a well-paying mechanic's job in Manila. Similarly, I described also in Chapter 4 the case of Danny, who had migrated to Manila several years earlier, but who had not been able to succeed and came back to work on his father-in-law's boat as a last resort or 'plan B'.

While it was young men who spoke to me about fishing within MPAs, and it was young men who are most obviously interested in the trappings of modernity, it would be potentially limiting to say that these are the only people interested in envisioning a future for themselves outside of poverty and fishing. McKay (2003), for example, has shown how older men and women can adopt alternative futures in Ifugao (in Northern Luzon) just as easily. In addition, there may be older people in the region who also engage in fishing within MPAs or other illegal fishing practices; they may be simply more discreet or modest about it. Notwithstanding these speculations, however, the limited data I have suggests that following Rigg (2006), it is younger people who more often adopt these practices. More strongly influenced by globalisation, younger people are expressing their desire to move out of the fishing sector for a variety of reasons.

Conclusion

Fishing within MPAs can thus be seen as a particularly bold and forceful variation on the discourse of the poor moral fisher. Fishers feel strongly about their rights to fish within all areas as long as they used 'legal' gears, and some are prepared to break laws regarding MPAs in order to do so. Because of the illegality of such actions, however, mostly only younger, bolder personalities attracted to higher risk are more comfortable with this behaviour. It is not a widespread practice, nor is it publicly acceptable. Indeed, it would be interesting to do further research on the potential inter-generational tensions caused by this form of fishing.

Whether the practice of MPA fishing is something that represents a broader, more significant long-term shift away from fishing in Coron, or a serious challenge to the otherwise dominant discourse of the poor moral fisher, is probably something that can only be completely answered with more long-term fieldwork over many years. It raises interesting questions about globalisation and the long-term future of fishing. Based on the evidence I gathered, however, I would argue that MPA fishing should instead be understood as a practice closely related to the discourse of the poor moral fisher, which is informed by the particular model of masculinity that young men in Coron engage with at this time of their lives. Many of these young men, I would expect, will go on to become more like the older man described earlier, who was more satisfied with his lot in life (whether this was in fishing or working in Manila).

It is young men mostly involved in MPA fishing primarily because they have different aims and motivations with regard to money and, by extension, their lives when compared with older men. Younger men are interested in the 'jackpot' that will bring them fast income. This income may give them opportunities to increase their status within their peer group, demonstrate to future partners that they have breadwinning power, and provide a basis for them to move into a different livelihood.

Fishing within MPAs can therefore be viewed as a form of and attitude towards fishing that lies somewhere between legal and illegal fishing. One element of the representations of fishing within MPAs is the insistence that it is not as damaging as the use of truly illegal activities such as cyanide and dynamite, and that fishing within MPAs is a rational and justified response from poor fishers to unfair regulations. These fishers see fishing within MPAs as necessary to maintain a livelihood in a poor country, while cyanide and dynamite are seen as much more sensitive. This element thus draws on aspects of the discourse of the poor moral fisher in its emphasis on an appeal to fairness and the 'right to survive' (Blanc-Szanton 1972). The second element of representations of fishing within MPAs that I have analysed in this chapter has links to a form of masculinity that emphasises bravery and alternative constructions of future livelihoods. Fishing within MPAs, I have suggested, is symptomatic of broader ideals among younger residents to demonstrate masculinity, and to move away from fishing and out of poverty.

Chapter 7 shifts the focus from MPAs to another form of regulation that was contested during the time of fieldwork—a set of measures to reform the live fish trade.

7. The Politics of Patronage and Live Fish Trade Regulation

Kung mayaman ka [if you are rich] you have all the favours. *Kung mahirap ka* [if you are poor] you have no justice (Resident of Coron town).

Pag may pera ka, hindi bawal [When you have money, it isn't banned] (Fisherman from Coron town).

It should be noted that the enactment of Provincial Ordinance No. 941 was not a straightforward policy process.... This process illustrates that policy making in natural resource management is quite complicated, and is tempered by the interplay of political and business dynamics (Pomeroy et al. 2008: 63-4).

Arguments about regulation of the live fish trade erupted through the last few months of 2006 throughout Palawan Province to an extent that is only hinted at in the quote by Pomeroy and colleagues. This chapter details how the process of regulation was understood, expressed and contested by different stakeholders. It shows how the discourse of the poor moral fisher was an important part of how the regulations were eventually rejected, and illustrates how this discourse is deeply connected to notions about personalised governance and patronage.

On one side of the argument, proponents of regulation such as conservationists and provincial level politicians called for a closed season, saying that such a move was necessary to protect the future of the industry. On the other side, opponents of regulation such as the live fish collectors, traders and municipal governments argued that such a move would have a massive impact on their livelihoods and greatly increase poverty in the region. The process of regulation was understood by local fishers as political and representative of a pattern in fisheries governance, and more broadly, of governance generally in the Philippines, that was understood as unjust and inequitable. I show how the process of implementing a closed season, the fishers' critique, and the subsequent overturning of the regulations exposed the way personalised politics is understood and practiced within Philippine society.

Firstly, I situate my argument with reference to other perspectives on fisheries governance. Secondly, I provide a background to the way in which the regulations were proposed and developed. Thirdly, I detail the manner in which fishers expressed their resentment at the regulations, and analyse this resentment in terms of local understandings of governance and governments. Finally, I discuss some of the implications of these understandings.

Approaches to Decentralised Fisheries Management and Governance

As I outlined in Chapter 5, decentralised approaches became a key feature of fisheries management in the Philippines in the 1990s. Recent critiques of these decentralised approaches to fisheries management have focused on the problems surrounding corruption, governance and participation, arguing that participation and genuine co-management have been limited by the lack of any honest attempt to devolve power to local communities (Walley 2004; Eder 2005; Lowe 2006; Ratner 2006). In Indonesia, Lowe (2006: 151–2) contends with regard to the management of the live fish trade that

> [w]hile poor people were the first to suffer penalties they also assumed the greatest risks, yet were excluded from the highest live fish profits. Laws, as they were written, interpreted, and enforced within the entrepreneurial Indonesian bureaucracy, enriched bureaucrats and their organizations and failed to protect either species or citizens.

Walley (2004: 54) relates how in Tanzania, local fishers stressed the need to crack down on corruption in the Maritime Division instead of the need for modernisation and development among fishers; in Cambodia, Ratner (2006: 84) argues that 'policy reforms must be complemented by institutional reforms, which are proving more difficult to achieve'. Elsewhere in Palawan, Eder (2005: 166–7) asserts that:

> [i]n the absence of more strenuous efforts to channel greater power and authority to the weakest and most poorly represented local groups, the efficacy of community participation will remain limited, and effective co-management regimes will prove difficult to institutionalise.

He shows how the burdens of community-based coastal resource management schemes were dispersed through ethnic, gender and class inequalities. What these arguments have in common is a concern with the ways in which fisheries management initiatives tend to get 'sucked up' into pre-existing social and political inequalities related to poor governance. From this perspective, the problems in fisheries management are due not to the poverty or ignorance of the fishers, but primarily to poor governance by local elites. As I shall show, this critique of governance corresponds with the argument made by local fishers in the Calamianes. My argument in this chapter aims to extend these ideas further in relation to the Philippines, showing how the development and subsequent demise of the live fish regulations reflects the distinctive nature of political patronage in the Philippines.

However, instead of simply declaring that poor governance from local elites is the sole or primary issue—as fishers in the Calamianes and some of these analysts suggest—I show in this chapter that fishers' demands for good governance are located within broader political practices and attitudes. I demonstrate how the practices of politicians at different levels of government, traders and fishers *all* operate by the same ethos of personalised politics that characterises the Philippines. Poor governance is certainly a valid critique of the fishers. However, through an ethnographic demonstration of how fishers understand and practice politics at multiple levels, I show how they in effect reproduce the very rules of the game of political patronage that they are critiquing.

Background and the Case for Regulation

Regulatory Framework for the Live Fish Trade in the Calamianes

Attempts to regulate the live fish trade in Palawan have a long history (Fabinyi and Dalabajan 2011). This is in large part because of the significance of the fishery. International attention has focused on the live fish trade since at least the 1990s, mostly attempting to stop the problem of cyanide fishing and the perceived 'boom and bust' nature of the trade (see Scales et al. 2006). The live fish trade in Palawan has expanded rapidly since the 1990s to become an important livelihood for many coastal communities across the province, and there has been a great deal of effort spent by policymakers on trying to regulate the fishery. The live fish trade has been particularly important in the Calamianes Islands (Fabinyi forthcoming); as such, attempts to reform the live fish trade are, together with the creation of MPAs, another major form of marine resource regulation in the region.

In his article on the problem of cyanide fishing in the Calamianes, Dalabajan (2005) lists various legal mechanisms dealing with a range of aspects of live fish trade management that have been developed since the late-1980s. These include the banning of compressors that were used in cyanide fishing, prohibitions on the use of noxious substances, and the empowerment of multiple agencies to enforce these laws. As indicated by the persistence of cyanide fishing, these laws appear to have had little efficacy. They are contradictory, highly confusing and are not enforced. As I showed in Chapters 4 and 6, the capacity for law enforcement is extremely low in the Calamianes. Dalabajan (2005: 4) points out that, because of the weakness of the law enforcement structures in the Calamianes, it is not surprising that in the last four years there have been no cases filed by law enforcement agencies in the Calamianes for violation of these provincial ordinances. Furthermore, the law is so confusing that some believe it was not intended to be enforced in the first place.

Since the attempts at regulation described in this chapter, efforts have focused on developing a quota system for the province (Fabinyi and Dalabajan 2011). At the time of writing implementation was ongoing and the outcomes were uncertain. Thus, despite their number and numerous aims, it can only be said that the laws have so far failed to capably regulate the live fish trade in Palawan.

Calls for Further Regulation

Much of the intellectual reasoning behind the calls for further and more tangible regulations were based on clear indicators that the Calamianes were essentially chronically overfished by 2006. A report by WWF stated in 2003 that:

> [t]he biological and ecological indicators suggest that the industry is 'mining' and degrading its resource base, greatly compromising its current and future regenerative capacity. Catch has lately been declining and any short-term increase in catch comes from fishing grounds outside the Calamianes.... The mean size of fish collected is getting smaller and exploitation rates indicate serious overfishing (Padilla et al. 2003: 8).

The report emphasised that by various measures—using ecological, economic and social indicators—the trade was extremely unsustainable. Another report from Conservation International (CI 2003) also highlighted the problems of cyanide and overfishing, calling for a stronger enforcement regime and networks of marine protected areas tied to the creation of alternative livelihoods for community members. Another report analysed the incapacity of the law enforcement structure to deal with the problem of cyanide fishing (Mayo-Anda et al. 2003). A policy brief commissioned by the Palawan Council for Sustainable Development (PCSD) in 2005 had also restated ongoing concerns over destructive fishing methods, overfishing and the targeting of spawning aggregations and immature fish (Pomeroy et al. 2005).

All these reports emphasised social factors as prime concerns for the sustainability of the live fish trade and any potential regulation of it. The WWF report for example acknowledged that

> [u]sing the environmental impact assessment framework, the logical recommendation would be to impose a moratorium to allow the fish stocks to recover. But from the sustainability assessment framework, which gives due consideration to economic and social impacts,

rationalization of the industry is recommended to address the problems of overfishing, cyanide fishing, and intrusion of migrant fishermen in to Palawan waters, among others (Padilla et al. 2003: 11).

Throughout this report, the authors emphasised the inequalities inherent in the industry, stressing the absence of empowering mechanisms for live fish collectors, the inequitable distribution of benefits and the growing sense of dependence (ibid.: 90–1). Similarly, the CI report cautioned against knee-jerk blanket bans of the trade, noting that:

> without alternative income earning opportunities, the short-term opportunity costs of MPA establishment in the Calamianes Region are likely to be higher than many fisher folk in these communities, particularly recent poor migrants, can initially afford. Suggestions on banning LRFT [live reef fish trade] without viable alternative income sources and effective enforcement only encourage more illegal activities and make them [the illegal activities] even harder to track down (CI 2003: 15).

The policy brief itself stated that '[t]he policy goal is for a sustainable fishing industry in Palawan Province that ensures viable fish stocks, ecosystems and livelihoods for present and future generations' (Pomeroy et al. 2005: 33). As I showed in Chapter 3, it too was concerned with the social impact of the live fish trade, arguing that the benefits of the trade were short-term only and were accompanied by many long-term costs. The writers of the report subsequently observed that '[s]uggestions to ban the trade or to regulate the fishery, which will impact upon the income of the fishers, will not be effective without viable alternative livelihoods' (ibid.: 36).

Thus, the calls for regulation of the trade can be seen as trying to address concerns not only over environmental degradation resulting from the trade, but also over the poverty of communities of the Calamianes. The writers of the reports were united in their views of the trade in its current form as something that was degrading the environment and increasing the risks of long-term poverty. Regulations as a response to this situation aimed to address both issues. The nature of these regulations, being grounded in social as well as environmental concerns, reflect the way conservation is often discussed in the Philippines.

Another significant, more general factor behind the push for regulation was Palawan's reputation as the 'last frontier' of the Philippines (Eder and Fernandez 1996), which I outlined in Chapter Two. The emphasis on Palawan as a green, clean place to live has ultimately meant that the discourse of sustainable development has gained considerable traction: many politicians have built their careers on being seen to adopt this discourse.

Thus, the various reports by NGOs and government agencies, the power of the green discourse in Palawan, and clearly the simple reality on the ground that there were many fewer fish than before were the prime factors behind the push for regulation. This push had a long history behind it, and Pomeroy et al. (2008: 63–4) detail some of the difficulties:

> It should be noted that the enactment of Provincial Ordinance No. 941 was not a straightforward policy process. Although a consensus was reached to adopt a regulated live reef food fish trade in June 2005, it was not readily accepted by all the provincial legislators. In fact, the development of the ordinance underwent several revisions, including a version which involved banning the LRFF (live reef fish for food) trade. The Provincial Ordinance No. 941 was finally enacted on 21 March 2006 and duly approved on 18 April 2006. This process illustrates that policy making in natural resource management is quite complicated, and is tempered by the interplay of political and business dynamics.

Pomeroy and his colleague's analysis of the policy process provides a good background to the events that took place, but the last sentence in the quote above is the only one which indicates some of the political dynamics that were vitally important to the ways the live fish regulations were received in the latter part of 2006. The rest of my analysis concentrates on how some of these 'political and business dynamics' affected the policy process after the regulations were approved at the provincial level.

Provincial Ordinance 941

In March 2006, Provincial Ordinance 941 (PO 941) was passed by the Provincial Council (Sangguniang Panlalawigan), entitled 'An ordinance providing sustainable and integrated regulation of the live reef fish industry, imposing certain conditions for the catching, trading and shipment of live reef fish out in the province, providing for violations hereof and for other purposes'. Provincial Ordinance 941 had many varied goals, including the creation of marine protected areas in specific spawning sites of groupers and overfished areas. However by far the most revolutionary part of the document, and the part that was subsequently highlighted by live fish collectors was Section Seven, which declared a closed season for the months of November–December and May–July. Unlike other previous regulations, this aspect of PO 941 would be relatively easy to enforce, as all commercial shipments of fish in the Calamianes have to be weighed and checked at the municipal BFAR office before transport. While the PO was passed in March 2006, the political system of the Philippines requires that provincial ordinances have to be approved by specific municipal ordinances before they are actually implemented in those specific municipalities.

Thus, for most of 2006, the provincial council and the various municipal councils that supported live fishing were locked in a stalemate, arguing the merits or otherwise of Provincial Ordinance 941.

At different points through 2006, the provincial council and the Palawan Council for Sustainable Development (PCSD) threatened the municipalities involved in live reef fishing with a moratorium on the renewal of all live reef fishing accreditations if these regulations were not adopted. Dinners and various meetings were held where the provincial council restated its aims of reforming the industry. At one meeting, the then vice-governor (and chairman of the provincial council) declared to a cluster of live fish traders: 'We are not killing the industry! If anyone will kill the industry it is the industry itself!'. He went on to argue that the municipalities were not 'respecting' the provincial ordinance by failing to create their own municipal ordinances.

Politicians at the provincial level framed their argument in terms of the environmental sustainability of the industry and Palawan more generally. At a meeting of the PCSD held in Coron, Congressman Mitra referred to the prominent fisheries researcher Daniel Pauly in his opening 'inspirational remarks'. Focusing on the live fish trade in these remarks, he went on to declare that:

> [w]e have always been guided by our mandate and commitment to ensure environmental sustainability by integrating sustainable development principles in all our programs and policies in Palawan. Let us put additional attention and effort to radically reverse the loss of our marine and coastal resources while time is still in our hands (132nd PCSD Regular Meeting, 25 August 2006, Coron, Palawan).

At one meeting between the Calamianes Live Fish Operators Association (CLOA) and various politicians, another prominent provincial politician implored the live fish traders to think about the province as a whole. He pointed out that many live fish catchers now had to go to the Palawan mainland to find their fish, and that Palawan as a whole only had room for 200 licensed live fish traders. He argued that '[t]he marine resources of Palawan can become infinite if we only care for them'.

Finally, on 12 December 2006, after a distinct lack of activity exhibited by the live fish traders associations and the municipal governments in creating a specific municipal ordinance, the provincial BFAR office enforced a ban on all live fish exports from Palawan. Immediately after this, furious lobbying and negotiations occurred. Fishers went to the traders that supported and financed their activities, asking for help. Many of the traders (numbers of whom were also municipal councillors) went to the provincial capital Puerto Princesa to lobby the provincial council to overturn the ban. In Coron, the mayor publicly declared at a Christmas party for live fish traders that the municipal government was totally committed to the live fish trade. He told the assembled traders that he had personally visited the governor of Palawan to try and persuade him to change the regulations. Throughout this period, rallying speeches were held outside the municipal hall, and municipal council members and traders left together for the capital as the fishers waited anxiously back in the municipalities and barangays. One of the most dramatic instances of this attracted the interest of national newspaper the *Manila Times*. It reported that:

> [s]ome 500 live fish gatherers and traders last week trooped to the capitol's legislative building in this city [Puerto Princesa], asking the provincial board to suspend the implementation of Provincial Ordinance 941, also known as the Palawan Live Reef Fish Ordinance. They claimed the ordinance … has made their lives difficult, especially this Christmas season (*Manila Times* 2006).

Dalabajan (2009: 57) later wrote of one meeting during this process that in 'a show of force, the traders' posse of lobbyists, largely composed of municipal legislators, descended to the capitol building'. Their arguments about the negative effect the ordinance would have on local fishers 'came with a veiled threat that this cruelty will be well remembered come election time, which was due in six months' (ibid.).

The provincial council 'melted' (ibid.), the ban was overturned at the start of the new year, and shipments were allowed to go through again.

The Case against Regulation

As described by fishers, the case against regulation of the live reef fish trade was similar to arguments against other regulations such as MPAs. Here, I focus more on broader concerns about corruption and inequality in fisheries governance. Fishers declared that regulations such as these punished all fishers for the activities of only a few, namely those who used illegal fishing methods. These regulations impoverished them, and ignored what they viewed as the real problem—the lack of enforcement against illegal fishing. The imposition of the regulations was seen to be ignorant of the needs and dignity of the poor and was viewed as an example of unjust governance. Here I detail the substance of these claims, firstly by looking at what people said about the live reef fish regulations, then by examining how fisheries governance more generally was perceived. Following this, I analyse these arguments with reference to broader conceptions of governance.

Attitudes towards Live Fish Regulation

At a dinner with representatives of the provincial council in August 2006, CLOA issued an appeal against the provincial ordinance on live fishing. The arguments it mustered in defence of the live fish industry were primarily grounded in how it had provided a strong injection of socio-economic benefits into the communities of the Calamianes. In a manifesto presented to members at the dinner, CLOA emphasised the increase in financial capacity that directly derived from live fishing. Live fish collectors were now 'able to start sending their children all the way to college and provide their families with a modicum of modern comfort like televisions, stereo and the like' (CLOA 2006: 1). If the live fish trade were to be banned or regulated, CLOA stressed, these poor fishermen would return to the life of even more extreme poverty that had marked their lives before the introduction of the trade. As one trader told me: 'Many people will go hungry with these new regulations'. These complaints were reiterated throughout all the interviews I conducted with the live fish traders, a Christmas dinner in which they restated their appeal for help to the municipal officials, and submitted through numerous representations to the provincial capital.

While the appeals of CLOA could easily be dismissed as a blatant attempt at using the vulnerability of live fish collectors to conceal their own motives of avoiding any regulations and maximising their own profits at any cost (one provincial council member identified their plea as such in his reply to their request), live fish collectors themselves also talked passionately about these issues. When discussing the ban during December, the same common responses could be heard repeatedly: the live fish trade had been a great help to the people of the Calamianes, there was no other alternative to live fishing during the windy season of amihan, and many people would go hungry because of the ban. People pronounced that there was no alternative livelihood. Also, the proposed ban for May to July would fall squarely in the middle of the most profitable season for live fishing with good weather conditions (habagat). The other proposed ban, from November to December, fell at a time where fishing was greatly decreased, and when fishers found it very hard to get by anyway because of the high winds and poor catch associated with this time of year. Many non-fishers registered their opposition as well, saying that the decreased income of live fish collectors would adversely affect other community income sources such as general stores.

A key issue that upset many live fish collectors was the way in which the regulations were broadcast. While these issues were being negotiated throughout 2006 at least, many fishermen were continuing to make investments in live grouper fishing, believing that this was the most prosperous fishery, and had a solid future. Fishers complained of a lack of a public hearing concerning the regulations, and while they were nominally represented in meetings such as that between CLOA and the provincial politicians, all the meetings I attended were numerically and substantively dominated by traders and politicians. Padilla et al. (2003: 87) have already drawn attention to this issue, noting that '[c]ommunities are essentially marginalized from the decision-making process, which is perceived to be controlled by local political and financial elites'.

One live grouper fisherman, Geronimo,[1] who had only recently changed from net-fishing to live fishing in the hope of accessing the 'jackpot' that live fishing could often bring, complained bitterly about the regulations:

> I've experienced two storms this week: Seniang [the name of a strong typhoon that hit the Calamianes just as the ban was enforced] and now this problem. I don't understand, I haven't been able to pay back any of my debt [his live fish operations were financed by a buyer], not even one cent. My livelihood is finished now.

As I bought some washing powder at the local *sari-sari* store later that day, he joked that I should bring my clothes back to him and he would do my washing, as he didn't have a job anymore.

What was striking about many of these discussions was how closely they were linked with a distrust of the motives behind the regulations. Almost universally, the fishers I spoke to during this period argued that the problem was solely about enforcement, not about the lack of regulations. As one angry fusilier fisherman put it: 'The problem is illegal fishing. This is the job of the police … the provincial government shouldn't be trying to stop live fishing completely … it is only some who are illegal fishers'. These fishers argued that the motivations behind the regulations were fundamentally part of political manoeuvrings that had little to do with any genuine conservation or management objectives. Many fishers suspiciously queried the timing of the implementation of the ban. One stated: 'They are introducing this now because it is just before Christmas. Now is the time they need to spend, to give money out to their voters. They [the provincial council] are just doing this to get money out of CLOA … *Mga politicos* [these politicians]—I know their style!' he told me, shaking his head. Many other fishers agreed with this interpretation.

A more complex hypothesis that gained a considerable amount of traction was that the ban was part of an elaborate scheme by the president to install an alternative governor of Palawan. According to this theory, the president had recently switched her support to the opposing candidate who was preparing to run for governor in the elections of 2007. The opposing candidate was hostile to the live fish trade continuing in the Calamianes because he was representing the allegedly jealous southern municipalities of Palawan, where the trade was

---

1 Geronimo was a fusilier-fishing captain during habagat but changed from net fishing to live grouper fishing during amihan over the time I lived in Esperanza.

not nearly as prominent (and in some instances banned). Although this may well have been the case, my point here is not to speculate about whether such assessments held any truth to them. Instead, I want to show how the deeply-felt cynicism about government behaviour informed understandings of live fish regulation.

From these descriptions of what many fishers saw as the motives behind regulation, it can be seen that some felt that the provincial government was going to be actively gaining through these regulations—either through extortion of money from CLOA or through political support from the southern municipalities. More importantly however, was that they were going to seriously damage the interests of the poor live fish collectors. In this way, live fish regulations were understood by many as something that would hurt the poor, and enrich the elites implementing them. In the next section, I explore in greater detail how broader patterns of fisheries governance were also viewed in this way.

Fisheries Governance

When I asked one older net fisherman about his thoughts on a new MPA being planned in the sitio next to the one in which he lived (San Andres), he gave a long sigh before launching into what he considered its impacts would be. He felt that the authorities should be concentrating on catching the illegal fishers and tightening up on corruption instead of increasing the restrictions on the 'small people' (*mga maliit*). He pointed out the negative impact that another MPA had already had on some of the fishers from his sitio (Esperanza), who had been required to find new fishing grounds. He didn't understand what the purpose was of punishing the small-scale fisher, while ignoring the real problem of illegal fishing. He described the 'style' of fisheries governance in the Philippines, where it was easy for illegal fishers to get away with their activities through simple bribery. If you get caught illegally fishing, he explained, you just pay a fee to the enforcing agents and they let you go on your way. For the poorer, small-time, legal fishers though, this was not an option.

A striking anecdote from fieldwork indicates how strongly some of the legal fishers felt about the problem of illegal fishing and the links to corruption. After several drinks at the local videoke house, Manuel, the fusilier-fishing captain whom I described in Chapter 6, started to talk about how punishment for illegal activities could easily be avoided with a bribe to the relevant authorities. Before he became a fishing captain, Manuel was a captain for a liveaboard diving boat. The owner was an Englishman who was addicted to an illegal type of methamphetamine (*shabu*) commonly used in the Philippines. Eventually the local policemen caught him with the substance and arrested him. After a quick negotiation, he paid them a bribe and avoided going to jail, being deported instead. Resentfully, Manuel mentioned that as I was rich in the Philippines, I would never have any problem if I ever got in trouble with the police.

Getting more worked up, he described what had happened on his last fishing trip. Travelling between his fishing grounds east of Palawan, he was suddenly rammed in the side by a Philippine National Police boat based in Coron, damaging the hull. Boarding immediately with shotguns at the ready and large aviator glasses on to cover their faces, they demanded to see the captain. They confronted Manuel and asked him whether he was using dynamite. After searching the boat and finding nothing, they unloaded two drums of fuel onto their boat. At ₱40 per litre, this added up to ₱24 000 worth of diesel—about US$480—a considerable expense for the owners of the boat. The police then helped themselves to a meal and took off with the fuel, leaving Manuel's battered boat behind.

Manuel became increasingly agitated as he talked about this, repeating over and over 'Kami [we], legal…. I am a legal fisherman'. Generalising from this experience, Manuel went on to talk about how that if you get rich through illegal fishing, it was easy to obtain protection (protektahan) from the police. He cited the example of one nearby island, where residents allegedly paid the police ₱20 000 (about US$400) every month to look the other way. If you were not rich and used legal methods however, you suffered. Manuel felt tremendously angry and hard-done by, because despite seeing the activities of dynamite fishermen all the time in his fishing grounds, he was the one who was targeted because he had no money (walang pera) to bribe the enforcement officials. 'Write that in your book—that is fishing in the Philippines!' he angrily commanded me.

As I have illustrated throughout this book, most fishers I worked with feel deeply disillusioned about the system of fisheries governance generally in the Philippines; they feel it empowers a few richer, illegal fishers who degrade the environment while penalising the legal, smaller fishers who are harmless to the environment. I now move on to a discussion of how these perceptions nest within perceptions about governance and corruption more generally in Philippine society.

Idealised Patrons and Corrupt Politics

I wish to focus here on what I understand as a significant disjuncture between poor Filipinos' conceptions of the ideal leader and their understandings of the more grubby reality of 'politics'. This will lead to my discussion of what I see as important links and continuities between these two forms of political leadership.

In its investigation of poor Filipinos' views of leadership, the Institute of Philippine Culture (IPC 2005: 19–42) emphasised what poor people consider to be the most important qualities and characteristic of a good leader and election candidate. These include being morally upright, dedicated to improving the lives of the poor, and impartiality. These characteristics were usually cited as belonging to local leaders. Conversely, participants in the study stressed corruption as the single most important factor in determining a bad leader (ibid.: 42–3). National officials were frequently cited as examples of bad leaders, and it was suggested by the IPC (Institute of Philippine Culture) that 'the perceived badness of national figures overshadows that of local leaders ... the participants find it hard to mention national figures as exemplars of good leadership' (ibid.: 47). Thus, pro-poor and impartial leaders at the local level are contrasted with corrupt and self-serving leaders at higher levels of politics.

I now develop these ideas with reference to my fieldwork experiences, by contrasting fishers' images of local leaders with their descriptions of politics more generally. From the local level, where politicians are usually conceived of as benevolent patrons, to the national level, whereby they are predominantly viewed as parasitic and corrupt, perceptions of politicians tend to be either of an idealised leader, or a crooked cheat. The behaviour and attitudes of poorer people correspond accordingly with these perceptions.

The captain of the barangay I lived in was held in what I saw as extremely high esteem by the citizens of my sitio, Esperanza. Regarded as a genuine 'man of the people', he was seen as someone who devoted his office to the care of his predominantly very poor barangay citizens. 'If you ever have any problems or need anything, go to him' I was advised when I first settled into the barangay. In particular, he was regarded highly by these citizens because of his role in supporting them in their protracted land dispute with the wealthy family that claimed to own the land they lived on. This consolidated his reputation for being pro-poor. As a prominent live fish trader himself, he was also viewed as someone who could represent the villagers and shared their concerns in the issue of live fish regulation.

The barangay captain's reputation for being pro-poor was an important factor behind his popularity. The need to look after the poor is commonly cited by members of Esperanza as the primary responsibility of governments, and, like elsewhere in the Philippines, the idea that well-off richer people should care for and accommodate the wishes of poorer members of society is well-ingrained. The 'right to survive' (Blanc-Szanton 1972) is a value that politicians in particular needed to be seen as respecting, and the live fish moratorium was seen as trespassing on this right.

An example of how local residents appreciated this concern for the poor could be seen when a petition of the barangay captain was brought around for people to sign. The petition asked for the support of the people in opposing new elections in 2007, arguing that the local government had not had enough time to fully implement their policies effectively. What was notable about this exercise, as the petition promoters moved from house to house, was the lack of any meaningful discussions about the implications of the proposal. Instead, once the people found out that it was a project of the barangay captain, they signed immediately. 'As long as the barangay captain helps us against that man [referring to their opponent in the land dispute], I will sign whatever he wants' one man declared. The assumption was that as long as they were receiving support from the captain in their land dispute, their loyalty to him and any of his projects would be guaranteed—no matter what the implications were. The enactment of the patron-client relationship between the barangay captain and the residents here illustrates the flipside of corruption—the provision of services to kin and loyal clients.

The barangay captain's popularity at the local level in Esperanza was virtually uncontested. While I didn't have sufficient data from other sitios to make a judgment on the extent to which his popularity was contested in other regions, the fact that he had been barangay captain for over a decade was an indication of his success. However, in other localities in the region, such political power at the local level was contested.

The metaphor of the parent is a common phenomenon when referring to 'good' local leaders. Dumont (1995) has documented the extensive use of kinship metaphors in political life in Siquijor Island in the Visayas. The metaphor of family was extensively used by politicians at various levels as a symbol of unification and togetherness:

> The kinship ideology stressed the natural character of kinship links, which entailed social solidarity, which in turn assured that smoothness prevailed in interpersonal relations. Such an ideology was so powerful and so pervasive that most political relations, whether, local, provincial, or national, were expressed through that idiom. Whenever the mayor spoke to his constituents, often with rhetorical flair and genuine eloquence, he emphasized the kinship links that bound them all (ibid.: 18).

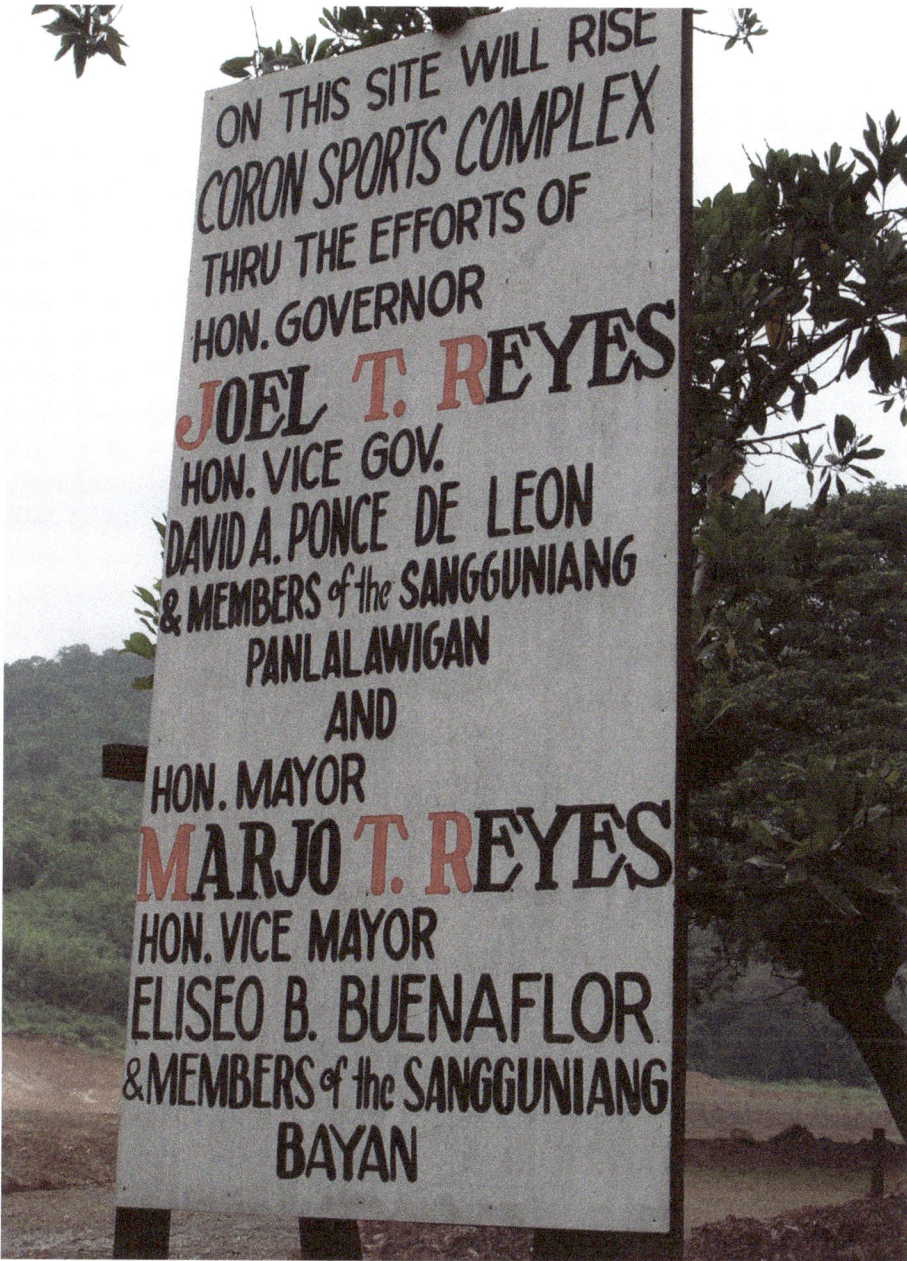

Plate 7-1: Infrastructure promotion like this is a common feature throughout the Philippines.

Similarly, the IPC report quoted earlier stated that '[p]articipants agree that a leader's position is comparable to that of a parent who acts as a guide and provider' (IPC 2005: 98). Local leaders in the Calamianes are often referred to by the metaphor of fictive kinship. 'I always refer to him as my father' stated one barangay citizen, as he described how he consulted the captain before making any major decisions. Both the local sitio chairman and the sitio representative in the Barangay Council were also referred to numerous times as father (*'tay*) or older brother (*kuya*) by residents who did not have actual familial relationships. Additionally, local politicians are often called upon to become godparents of the children of poorer families. Politicians thus become drawn into fictive kin networks of poorer families. These relations of kin may be fictive but they have real resource implications. Politicians are expected to provide support at occasions such as weddings, and help those members of the fictive kin network when needed. One councillor related to me how an elderly woman from his barangay approached him shortly before the municipal elections, asking for his help in acquiring a new set of false teeth: 'I would like a new set of dentures so I can smile when I vote for you', she told him. Another politician recounted his frustration at his inability to govern in an impartial way, because of the perception among his constituents that the responsibility of government was to look after them 'from womb to tomb'. In this way, the state becomes personalised as a patron to whom one can appeal to for pity.

Comments about politicians at the municipal level usually depended on whether the person I was talking to had personal ties to the politician. One member of the municipal council whose father used to employ most of the fishers in Esperanza, for example, was extremely well regarded. The majority of the other councillors were frequently simply labelled as 'corrupt'. Instead of allegations against specific municipal politicians, however, complaints about corrupt politics were more general. Thus, when people complained about the behaviour of politicians, their complaints were usually not directed against politicians that they knew, but against politicians at more remote levels whom they did not necessarily know.

Like the authors of the IPC report, I found that most common were complaints relating to politicians at the national level. Informants frequently told me how the system of politics was deeply corrupt in the Philippines, some concluding that things were better under the Marcos years because then at least there was 'discipline'. This system of politics is understood as enriching the lives of the politicians while taking advantage of and living off the hard work of poor people like themselves. Corruption is a wearily unloved and assumed characteristic of government generally in the Philippines; 'this doesn't happen in your country, Mike' (*wala sa inyo* Mike) my informants would frequently tell me after every such conversation. As one fisher put it when talking about the alleged links between the traders and municipal councillors: 'they are very opportunistic (*mapagsamantala*)'. The IPC (2005: 46–7) report states that being corrupt was unanimously nominated by the participants as a distinctive quality of being a bad leader; 'corruption is clearly detested....' The ways in which people hear about issues of corruption, primarily through sensationalist media reports and the everpresent gossip (*chismis*), reinforces this perception of politics. The perceived links between corruption and poverty are highlighted by 2010 Presidential candidate Benigno Aquino's successful use of the campaign slogan '*Kung walang* corrupt, *walang mahirap* [If there are no corrupt people, there are no poor people]'.

My point here is that the live fish trade regulations fitted in with a pattern of fisheries governance and general governance that is understood as essentially anti-poor and corrupt. The regulations were established at a level where corruption was understood to be the mode of operation generally among politicians to whom people had no personal connections; how the regulations were broadcast so suddenly and with minimal public consultation suggested a lack of transparency and self-serving elitism; and the massive impact they threatened to have on the livelihoods of local fishers indicated a severe deficiency in caring about the poor. The regulations therefore became consigned in peoples' minds to the domain of 'bad politics' or 'corruption'. Responses among poor fishers reflected this cynicism. Rightly or wrongly, PO 941 became linked in fishers' minds with a system of governance that fed on the misery of the poor.

Patronage and Corruption

From these accounts it can be seen that there is a wide gap between the idealised portraits of 'good leaders', typified by some local politicians, and 'politics' more generally, viewed through the lens of corruption. At the local scale, where poorer people may have links of fictive kin with certain politicians, the expectation is that the politicians will provide goods and services to these poorer people. At higher scales, where local people may not have links and therefore will not be receiving any benefits, the same practices are simply viewed as corruption. Local politicians such as the barangay captain who distributed benefits and resources are not understood as corrupt by local people, whereas the same activities of politicians at a higher scale are condemned as corrupt because they are not seen as looking after the interests of the poor. For the local people I worked with, the same sorts of activities are therefore understood in very different ways, depending on whether the patronage is directed towards them or not, and at what scale it occurs.

These understandings of political behaviour correspond with the argument made by Russell (1997: 91), who argues that 'many Filipinos expect state officials to use their resources to support their particular economic interests in return for voter support'. Citing Olive (1996), she asserts that 'many people hope or expect their local officials will favour their resource practices rather than neutrally enforce an abstract set of state regulations' (Russell 1997: 91).

In his analysis of public life in the Philippines, Mulder (1997) claims that social life exists in three arenas—family, community and impersonal society. Family life is the moral core of life, symbolised by the mother and concerned with intra-group relations (Mulder 1997: 17–28). Local politicians with whom fishers have personal connections are brought into this arena. At the other end of the spectrum however—the end where the idea of 'politics' is consigned by my fishing informants—is the public space of wider society, 'an area of opportunity where amoral power reigns supreme' (ibid.: 133). It is in this space where it is understood that crooked politicians make their shady deals, buy their votes with their 'goons, guns and gold', and siphon off public funds for personal wellbeing. As Mulder argues, '[t]he best thing little people can do is to attach themselves to such bosses who seemingly know their way around in that vast world, to seek their patronage, so personalizing access to "public" space without it becoming any of their business' (ibid.: 132). 'Politics is less about good governance and more about benefits' (Mulder 2000: 186). This analysis concurs with that of Pertierra, who agrees that 'in the absence of trust in appropriate public institutions, strangers can only be assimilated by their conversion into consociates.... Filipinos personalize the public sphere and where possible use its resources to pursue private gain....' (Pertierra 2002: 88–9).

Fishers characterised the relationship between themselves and the politicians at a higher scale trying to implement the live reef fish regulations as a form of 'negative reciprocity' (Sahlins 1972). The assumption was never that the state was going to manage the resources in a sensible and responsible manner; instead, fishers believed that the politicians would simply mine the resources for their own purposes or those of their clients.

Some of Mulder's argument can be criticised for demarcating too clearly or crudely the differences between family, community and public life, and patently not 'all' politicians and politics are consigned to the amoral public sphere, as Kerkvliet, among others, has pointed out (Kerkvliet and Mojares 1991). But it does illustrate important aspects of how politics is understood in the worldview of many Filipinos.

Conclusion

The failed process of regulation highlights several significant social realities in the Philippines. Corruption—or at the least poor governance—is certainly present to varying degrees in fisheries management in the Calamianes, and in this respect my paper agrees with the arguments made by Eder (2005), Ratner (2006) and others who assert that few inroads will be made in sustainable fisheries management until underlying social and political problems of governance are addressed. The opposition of fishers sprang primarily not from ignorance, but from cynicism about the political process of regulation itself. Other attempts at regulation had typically been applied unfairly, disadvantaging fishers. However, I have argued that the situation was about more than just poor governance. I have shown how these views of poor governance were embedded within particular understandings and expectations of political behaviour and patronage. The regulations were opposed and failed not only because of the poor record of other attempts at fisheries governance, but because they did not conform to peoples' expectations about political behaviour. The ways in which fishers conceptualised the regulations as a form of negative reciprocity is reflective of the way politics is understood and frequently conducted in the Philippines—it is about supporting particular groups of loyal supporters. Fishers used the discourse of corruption as a way of condemning a piece of legislation that did not provide any foreseeable benefits to them. United with their patrons, the fish traders and their municipal political allies, they used this discourse to force the provincial politicians to drop the moratorium. While their critique of corruption was certainly a valid one, fishers failed to recognise that their political expectations and behaviour exemplified the classic games of patron-client ties and political patronage that were being critiqued.

This chapter has shown how a piece of legislation that was driven by social and ecological imperatives could be turned on its head and labelled as corrupt, because it was not seen as delivering the benefits expected from a good patron in a moral economy. I am not arguing here that this is the way politics is always understood or turns out in the Philippines; Kerkvliet, among others, (Kerkvliet 1990, 1995, 1996; Kerkvliet and Mojares 1991) has demonstrated that this is often not the case, showing that elections, for example, are sites of multiple meanings. Indeed, the fact that the legislation managed to get as far as it did, and the ways in which corruption was used as a discourse, could also be used as evidence of this ambiguity. Dualistic frameworks that stress the 'traditional' understandings of poor Filipinos, as opposed to 'modern' (for example Alejo et al. 1996), are unable to account for this. But the chronicle of the live fish regulations does demonstrate how deeply entrenched these understandings of political processes remain among many Filipinos, and the diverse ways in which they are represented at different levels of government.

8. Conclusion

This book has sought to demonstrate how local fishers in the Calamianes Islands understand the relationship between poverty and the environment, and how those understandings have framed and contributed to the outcomes of marine resource management policies. To support this argument, I have aimed to explicitly link two, related sets of literature. By using the insights of others interested in understanding social relations, reciprocity and cultural patterns in the Philippines, the aim of my book is to highlight the cultural perspective in analysing Philippine environmental politics. There are thus at least three primary themes running through the work. Firstly, I have presented an ethnographic perspective on the poverty-environment relationship. Flowing on from this central theme, I have shown how such a perspective illuminates certain ideas about power and reciprocal relationships in the Philippines, and also how it affects environmental management. In this concluding chapter I analyse these three broad themes in the context of their policy implications.

In contrast to analyses that have concentrated on the objective features of the poverty-environment relationship, I have instead provided an ethnographic account of how some of the subjective features are understood, represented and enacted by local fishers in the Calamianes. As I demonstrated in Chapter 4, fishers perceive their relationship with the marine environment to be intimately connected to their own poverty. In this representation, their legal fishing practices are linked with the reproduction of poverty and a minimal impact on the environment; illegal fishing, conversely, is seen as responsible for all environmental degradation while providing a means to accumulate wealth. Illegal fishing and its ability to flourish, (as I showed in Chapter 7), is strongly linked to perceptions that the political elite of the Philippines is corruptly encouraging this behaviour. The poverty-environment relationship is therefore viewed as a strongly political relationship: if we do the right thing and fish legally, we remain poor, the fishers are saying, while those who do the wrong thing and use illegal methods are rewarded. In this way the perspective of fishers closely resembles that of the 'materialist' form of political ecology (outlined in Chapter 1), which tends to argue that poverty is not the primary factor behind environmental degradation.

Nevertheless, conservation organisations in the Calamianes tend to focus much (but not all) of their work on the activities of poor people. While these conservation organisations may not have explicitly subscribed to the 'vicious downwards spiral' model of the Brundtland Report that I introduced in Chapter 1, in practice they have acted upon the implied need to alleviate poverty in order to achieve better environmental outcomes in several ways.

One of the rationales behind the SEMP-NP project, for example, was to provide incentives for poor fishers to stop their presumably harmful fishing practices through the implementation of tourism user fees. The broader partnerships that conservationists were seeking to forge with the tourist industry were in large part justified by the implied need to persuade fishers to adopt an environmentally sustainable alternative livelihood. If fishers can afford to do something else, according to this logic, the pressure that they are placing on marine ecosystems will ease. Although conservation organisations were also trying to reduce illegal fishing and to lessen the impacts of commercial fishing in the Calamianes, they were still acting on their view that poverty and poor fishers were to a considerable extent responsible for environmental degradation.

Such different assumptions about the relationship between poverty and the environment show why the implementation of regulations played out in the way that I described in Chapters Five to Seven. Conservation organisations were acting on a model of the poverty-environment relationship that fishers did not agree with—namely, that poverty and poor people are responsible (even if not exclusively) for environmental degradation. Because of this, they were always going to face difficulties in dealing with the alternative model of environmental degradation espoused by fishers. In this way, I have argued for the value of adopting a cultural and discursive approach when trying to understand the local practices of environmental politics. It has not been my intent in this book to definitively argue whether or to what extent poverty is a driver of environmental degradation in the Calamianes, although I have suggested that environmental degradation was at least partly due to the legal actions of poor fishers. Instead, I have shown how differential constructions of this relationship have informed behaviour that shaped environmental outcomes.

Local Discourses: Their Context and Impact

'Asking for Pity'

A second central theme of the book has been to demonstrate how local conceptualisations of the poverty-environment relationship among fishers resonate with certain notions prevalent among poor people in the Philippines. Such notions, as analysed by Cannell (1999), Kerkvliet (1990), Blanc-Szanton (1972) and others throughout the country, articulate an ethic of fairness and justice for the poor. In the case of Bicol, for example, Cannell (1999: 228) writes:

> The people 'who have nothing' must spend a great deal of time in trying to 'disponir' [borrow money], that is, not only in borrowing money, but also more broadly in convincing other people with more resources and

more power to make available some of what they have, to recognise the existence and the humanity of people poorer and less powerful than themselves.

For Cannell, the articulation of this ethic is a reflection of the broader Bicolano view of power:

> … Bicolanos tend to see all relationships as dynamic engagements. There are people "who have nothing" and those who have wealth, beauty and power, but there is always potential for negotiation and persuasion, through which the painful gap between the two may be lessened, and the power-deficit of the poor, not eliminated but ameliorated (ibid.: 229).

Similarly, Kerkvliet (1990: 251) discusses the articulation of this ethic in the context of 'everyday politics', where poorer people make claims on richer people in a variety of ways. He argues that on the basis of the ethic of justice and fairness for the poor, poor people in San Ricardo 'frequently make claims on how resources should be used, distributed, and produced'. From this perspective, poor people here 'are neither passive nor mystified' (ibid.: 262). Instead, Kerkvliet interprets poor peoples' actions as a form of resistance against those more powerful than themselves. As Cannell (1999: 227) explains it, the stories told to her by Bicolano people were not told 'pathetically'; rather, they were symptomatic of the desire of poor people to change their relationships with richer people for the better. For Cannell, the fact that this cultural construction of power is so widespread and so broadly consistent in the lowland Philippines— 'including both "animist" and Muslim regions' (ibid.: 233)—suggests that it is a coherent approach to power more generally in the Philippines. Patron-client relationships are common in many other poor countries, such as Indonesia. However, the emphasis here on morality may not be found in Indonesia in the same form or to the same extent, and may reflect the pervasive influence of Catholicism in the Philippines.

Certainly, this 'basic rights discourse' and the value of fairness for the poor are not the only values expressed by or about poor people in the Philippines. I have also described various instances in the book where richer people in the Philippines have expressed contempt and fear towards poorer people. Indeed, these discourses are often intertwined (Clarke and Sison 2003). As Pinches (2008) puts it, '[u]rban poor housing in the Philippines is variously looked upon by outsiders with revulsion, fear or sympathy'. The point I make, however, is that the discourse of morality towards the poor is one which poor people will appeal to in particular when faced with hardship.

As I described in Chapter 3, a large part of the everyday economic relations in Esperanza are structured around the notion that people make claims on relatively richer people—within Esperanza, in Coron town, Manila and elsewhere—for assistance in various ways. Asking a fish trader to finance one's fishing operations when moving to a new fishery, or for a seat as a crew member on a commercial boat, or help for school fees for a child or for medical help for someone who is sick: all these are examples of poorer people making claims on those with more resources in the context of 'everyday politics'. 'Asking for pity' in this context is not viewed in the same way as perhaps the English expression suggests. As Kerkvliet (1990: 263) argues:

> Many villagers deliberately try to attach themselves to richer, more influential individuals in order to improve their chances of dealing with realities they must face. In this way, they purposefully put themselves in deferential relationships. But that does not necessarily mean that they are void of ideas and beliefs, other than those underlying deference, to guide their social actions.

I have argued that the discourse of the poor moral fisher among fishers in the Calamianes should be considered in a similar light. Without trying to impart any artificial coherence to the discourse of the poor moral fisher, or to adopt a romanticised view of this form of resistance, I have demonstrated that fishers look for fairness and justice with regard to marine resource regulation. Their argument that they are not responsible for environmental degradation, and that regulations should not apply to them, can be understood within this context of 'asking for pity' and of making claims on those with more resources. Here, fishers attempt to address inequitable social relations between themselves and those more powerful—namely the government, illegal fishers and those in the tourism industry. Presenting themselves as pitiful in the context of marine resource use and regulation, I argue, is an attempt to reframe the social world of fishers in the Calamianes in a more favourable way. Marine protected areas (MPAs) have been seen as chances to address some of the class relationships with the richer tourism industry, while resistance to the live fish trade regulations has also been driven by a sense that the role of government should be to support and nurture the weaker members of society at critical times. The responses of fishers should therefore not be seen *only* as short-term strategic resistance to environmental regulations, but also as encompassing long-term goals of addressing particular social relationships. Through the discourse of the poor moral fisher, fishers are trying to draw those with more resources into personal exchange relationships.

While the resistance of fishers to regulations in the Calamianes has similarities to the discourses of marginalised fisherfolk in other countries (for example Kurlansky 1999: 219-33), the specific idioms in which it is expressed and the cultural forms in which it is embedded are particular to the Philippines— especially the emphasis on morality and fairness for the poor. I have therefore highlighted the role of culture in this discourse throughout the book, and tried to avoid characterising it primarily in terms of class relations. Similarly, Cannell argues that a purely economic or capitalist explanation for the cultural idioms of pity she described is implausible, given the 'continuity ... with ways of viewing "oppression" in the lowlands historically' (1999: 239). In this sense, responses of fishers to marine resource regulations can be understood less as an example of straightforward class-based resistance, or as a social movement converging around relatively abstract notions of social justice, democracy, or rights as citizens of the Philippine state. Elements of these broader ideas are certainly present, and the work of activist and social justice oriented NGOs in the Philippines (Austin 2003; Bryant 2005) has done much to promote awareness of them among fisherfolk. Similarly, the ways in which fishers use the discourse of 'corruption' (as in Chapter 7) suggests that this is not just about the 'traditional' understandings of Filipinos. However, as I also illustrated in Chapter 7, even when fisherfolk appeal to the state, they tend to do so using the idioms of pity and through the particular patrons with whom they maintain reciprocal relationships. This illustrates how the understandings of social life, reciprocity and governance that I have focused on in this book are still a central part of life for many rural Filipinos.

Obviously, resistance based on promoting an ideal of justice for the poor does not always work. The ways in which the discourse of the poor moral fisher is able to produce beneficial outcomes for fishers appears to depend significantly on the ability of people to establish and maintain reciprocal social relationships with those in greater power. Whether the poorer inland farmers of the Calamianes, with their lack of personal connections to more powerful or rich people in town, could achieve the same sort of outcomes as the fishers is doubtful. Similarly, the discourse of the poor moral fisher does not always produce beneficial outcomes for everyone. In this sense, the discourse of the poor moral fisher is not simply a 'noble' form of resistance to be romanticised (Ortner 1995). As I illustrated in Chapter 5, for example, there are some residents who expressed enthusiasm for the creation of MPAs that would disadvantage their neighbours. In this way, those who are unable to form the social relationships that legitimate them as 'poor' may suffer as a result.

However, the dominance and relative coherence of this discourse is perhaps best explained with reference to Peluso's (1992: 11) analysis of 'village solidarity' among Javanese peasants who resist the state in the context of forest regulation. As Peluso argues, when the enforcement of the state's law impinges upon the 'moral economy' (Scott 1976) of a group of people, 'even a highly differentiated peasantry can mask its class tensions, imparting a Chayanovian solidarity to a normally strained set of social relations' Peluso's (1992: 11). Similarly, while there are different social classes and economic interests in Esperanza, the perspectives of residents tend to converge around the discourse of the poor moral fisher.

Richer fusilier and fresh grouper boat owners were not as affected by the introduction of MPAs because they fished elsewhere, nor by the moratorium on the live fish trade because they were not involved in this fishery. However, many of their crew members were certainly affected by both of these forms of regulation, as they may have fished part-time in the area covered by the MPAs or fished for live grouper during amihan. And, while commercial boat owners themselves may not have been affected as much, often the sons (or the close kin) of a boat owning family were involved in the fisheries that were more affected by regulation. Along with the more general idea that fishing in the Philippines (even as a commercial boat owner) is an extremely low status occupation, this provides additional reasons for boat owners to adopt the discourse of the poor moral fisher. And as I have shown throughout the book, an important component of this discourse is about the obligations of the richer to ensure the wellbeing of those with fewer resources. Richer boat owners in Esperanza therefore have strong material and social motives to support the claims of poorer fishers.

Richer residents outside of Esperanza and others who are typically subject to claims by the discourse of the poor moral fisher—such as the politicians in town who often have to make decisions about whether to punish infractions over incursions into MPAs—also have strong reasons to support and accommodate many of the claims. The discourse offers them an opportunity to be virtuous and respectable, and to gain the support of poorer residents—this is something that is particularly important for those who are involved in local politics. The claims of the poor moral fisher are about ensuring that these people who are more well-off continue to be engaged in a mutually beneficial social relationship. When richer residents try to remove themselves and deny such claims, the potential for conflict occurs (Kerkvliet 1990).

Impacts on Environmental Management

As I have shown through Chapters 5 to 7 in particular, the articulation and enactment of this vision for fairness among fishers is extremely important for the outcomes of marine resource management policy in the Calamianes. In Chapter 5, I illustrated the way in which fishers presented their views during debates about the construction of MPAs. Arguing for MPAs that supported the livelihoods of fishers, while taxing the practices of the richer dive tourism sector, fishers here represented the debates as arguments over social fairness. During these planning processes, in many cases fishers were able to incorporate their concerns into the MPAs. As a result, many MPAs in the Calamianes during 2006–07 had smaller core zones than originally planned and were adapted to the locations of fish traps of local residents, buffer zones were adapted to allow the use of many local gears, and enforcement was often lax when it was conducted by local communities.

In this context of lax enforcement, some younger fishers were willing and able to fish within MPA boundaries. In Chapter 6 I detailed how these fishers felt about fishing within MPAs—they believed it was their right to continue to do so, despite the regulations. Such younger fishers felt this way, I demonstrated, because of their particular economic and personal values. The enactment of their visions for the future has had a significant impact on the integrity of many of the MPAs in the Calamianes.

In Chapter 7, I argued that the discourse of the poor moral fisher was also a vital part of the general uproar surrounding the imposition of a moratorium on live fish trading and a series of associated regulations. By appealing to relationships with their wealthier allies, the live fish trading municipal elite, fishers were able to successfully participate in a movement that eventually overturned the moratorium and the essential elements of the regulations. Through these examples of MPAs and the live fish regulations, I have shown the distinctive ways in which marginalised people in the Philippines resist and reframe conservation regulations. In both instances, members of the local government and other key power brokers are typically pressured by social obligations to minimise the impact on local residents, and instead are invited to take part in reciprocal relationships. Because of the impact the discourse of the poor moral fisher has had on the politics of environmental management in the Calamianes, it is worth considering briefly some of the broader implications for policy that have arisen from this study.

Policy Implications

Lastly, I return to the theme of moral difficulty faced by managers that I mentioned at the beginning of the book. As this book has demonstrated, there are no clear or easy solutions to the problems related to marine resource use in the Philippines. From the perspective of the fishers in Esperanza, the 'political economy' of being a poor moral fisher is highly unjust—illegal fishers degrade the environment while getting rich, while poor legal fishers like themselves suffer and live in poverty. As I have shown, this claim is powerfully and successfully articulated in various contexts, in which a 'better deal' for fishers is actively pursued.

As the fishers would see it, these arguments have a moral force lent to them partly because of their poverty. In addition to this ethical question of the need to address poverty, the basic injustice of the issues as the fishers present them is stark. Consider, for example, MPAs. In a context where great damage was being done to the marine environment by outsiders fishing illegally, why was the response of the government to introduce regulations on fishing that restricted the practices of the 'small people' (*mga maliit*) who had a negligible impact? In the case of the live fish regulations, again it appeared to fishers as if government was ignoring the real problem of cyanide fishing while trying to punish the activities of other hard-working legal fishers. As Li argues with regard to development interventions more broadly, for these fishers, such attempts at regulation would have been seen as illustrating the 'conceit of a will to improve that directs the conduct of "small people" while leaving radical political-economic inequalities unaddressed' (Li 2007: 282).

On the other side of the argument, the need for some form of regulation is clear. As the comprehensive report from BFAR made clear (BFAR 2004), the fishery sector in the Philippines is in deep trouble. In the Calamianes, reports since the late-1990s by a range of national and international organisations have concluded that the live fish trade is unsustainable, and that there is a great need to better manage the various island fisheries. From this perspective, regulations to reform the live fish trade are the painful but necessary action required if the Calamianes are to maintain a future that continues to use marine resources in a productive and sustainable manner. Similarly, MPAs are necessary in certain vital coastal regions of the Calamianes where fish reproduce. From this perspective, simply following the desires of local fishers—who, despite the moral force that is lent their arguments because of their poverty, are still just one sector among other competing voices—and avoiding any form of regulation will ultimately spell disaster, not only for others with interests in marine resources, but also for the fishers themselves.

Tourism is another force driving change and regulation whose interests have to be recognised. Marketed as the 'next big thing', tourism operations in Coron have been competing directly with the fishing sector for access to coral reef ecosystems, and aim to create an economy relying less on resource extraction and more on leisure based on these resources. From this perspective, fishing has to be regulated strongly in order to make room for non-extractive resource uses.

The question then is whether it is possible to reframe the sorts of regulation that are clearly required in a way that acknowledges the social and moral claims of fishers in a more tangible way. While the regulation of marine resource use has long had to incorporate people in its management (especially in the Philippines), a more explicit focus on the moral claims of particular people may point the way to better and more equitable forms of management.

Recently, a great deal of academic interest has concentrated on resolving these issues of contesting interests over marine resources with reference to linked 'social-ecological systems' and 'resilience' (for example Berkes et al. 2003). Many analysts have argued strongly for the principles of 'good governance' that promote resilience, as articulated through the themes of adaptive co-management, polycentric governance, interactive governance and other related terms (Armitage 2008). Lebel et al. (2006), for example, hypothesise that the existence of participation, polycentric and multilayered institutions, and accountable and just authorities can be associated with an increase in resilience for social-ecological systems.

Authors such as these have lauded the potential of good governance to promote resilience and create a situation that will be mutually beneficial to all stakeholders. However, Armitage makes the crucial point that governance attributes such as those described by Lebel et al. (2006) are 'productive and important, but they are circumscribed by context, and ... provide only partial direction for governance innovation'. Citing Li (2007), he argues that they represent a set of '"prescribed" and normative governance values or principles' (Armitage 2008: 18). In contrast, perhaps what we need to be more aware of is the inevitability of 'hard choices' in marine resource management (Bailey and Jentoft 1990; McShane et al. 2011). Resilience, like any other concept for ecological management, including those that emphasise good governance— and the policies advocated by the managerial approach to marine resource management in the Philippines that I outlined in Chapter 1—cannot remove the need for political decisions and negotiations among diverse stakeholders to be made in particular local contexts. Any model of institutional arrangements will inevitably be infused with, and have to take account of, local political, economic and social contexts (Flyvbjerg 2001; McCarthy 2006). In many ways this is where the value of ethnographic description lies. Li (2007) has shown how the ethnographic method of asking, 'unlike the managed participation of the Nature Conservancy or the World Bank, is not constrained by the need to devise technical interventions' (ibid.: 280). Instead, with its focus on local processes and practices, ethnographic description can highlight real-world complexities of everyday life in a way that broader generalised models cannot.

The experience in the Calamianes shows clearly how attempts at regulation have been experienced by different groups in different ways, meaning that some groups of people have inevitably felt marginalised. In the case of MPAs, for example, I have argued that the focus on satisfying dive tourism operators, conservationists and fishers all at once did not occur. Instead, a focus should be to either compromise between the different aims and motivations that different sectors bring to MPAs, or to be more explicit about the role of particular MPAs—to define which particular sector the MPA is meant to engage with and satisfy—and go about improving the situation of those worse off in other ways. In the case of the live fish regulations, while proponents of the regulations argued cogently that such reforms were needed to protect both the ecosystems and the communities who used those ecosystems, they failed to allay justifiable fears about the dramatic short-term impact the moratorium and the subsequent regulations would have on live fish collectors. In the words of Li (2007: 11), these interventions were framed as a 'will to improve', and excluded 'political-economic questions—questions about control over the means of production, and the structures of law and force that support systemic inequalities'. They also excluded the cultural and moral perspectives that I have focused on in this book.

With a similar stress on grounded, 'real-world' complexities, practitioners in the field of integrated conservation and development projects have recently emphasised the need for greater attention to and understanding the goals and interests of all stakeholders around protected areas. They note that win-win solutions are usually unachievable: 'Once these different interests are identified, clarified, and understood', they argue, 'the opportunities for negotiation and tradeoffs can be explored' (Wells et al. 2004: 412; see also McShane et al. 2011). Such a focus on trade-offs does not necessarily make the very real problems of marine resource management that much easier. But it does provide perhaps a more useful way of looking at these issues. Instead of trying to provide solutions that will benefit all people equally, the careful study of particular contexts and interests may serve to highlight the issues of who will win, who will lose, and how to go about dealing with those outcomes.

In this book, I have contended that it is necessary to pay more attention to the interests of fishers with regard to marine resource use. Understanding these interests in greater depth from the outset may help at least to clarify how the process of negotiation over tradeoffs may occur. These perspectives of fishers, I have argued, are deeply entwined with particular locally conceptualised ideas about morality and the relationships of humans with their environment. Local people desire and expect to be treated with humanity, dignity and fairness, and policy that does not engage with these expectations in an explicit way risks being ineffective.

References

Acheson, J.M., 1981. 'Anthropology of Fishing.' *Annual Review of Anthropology* 10: 275–316.

———, 1985. 'Social Organization of the Maine Lobster Market.' In S. Plattner (ed.), *Markets and Marketing*. Lanham (MD): Society for Economic Anthropology and University Press of America.

Agpalo, R.E., 1972. *The Political Elite and the People: A Study of Politics in Occidental Mindoro*. Manila: University of the Philippines, College of Public Administration.

Agrawal, A., 2005. *Environmentality: Technologies of Government and the Making of Subjects*. Durham (NC): Duke University Press.

Aguilar, F.V., 1998. *Clash of Spirits: The History of Power and Sugar Planter Hegemony on a Visayan Island*. Honolulu: University of Hawai'i Press.

Alcala, A.C. and G.R. Russ, 2006. 'No-Take Marine Reserves and Reef Fisheries Management in the Philippines: A New People Power Revolution.' *Ambio* 35: 245–254.

Alejo, A.E., 2000. *Generating Energies in Mount Apo: Cultural Politics in a Contested Environment*. Quezon City: Ateneo de Manila University Press.

Alejo, M., M.E Rivera and N.I Valencia, 1996. *[De]scribing Elections: A Study of Elections in the Lifeworld of San Isidro*. Quezon City: Institute for Popular Democracy.

Alexander, J. and P. Alexander, 1991. 'What's a Fair Price? Price-Setting and Trading Partnerships in Javanese Markets.' *Man* 26: 493–512.

Anda, R., 2008a. '3 Vietnamese Fishers Nabbed in Palawan Town.' *Philippine Daily Inquirer*, 19 May.

———, 2008b. '14 Vietnamese Caught Poaching in Palawan.' *Philippine Daily Inquirer*, 31 May.

Arin, T. and R.A. Kramer, 2002. 'Divers' Willingness to Pay to Visit Marine Sanctuaries: An Exploratory Study.' *Ocean & Coastal Management* 45: 171–183.

Armitage, D., 2008. 'Governance and the Commons in a Multi-Level World.' *International Journal of the Commons* 2: 7–32.

Arroyo, G.M.A., 2007. 'State of the Nation Address.' Viewed 26 July 2007 at http://halalan-2007.blogspot.com/2007/07/president-arroyo-state-of-nation.html

Austin, R., 2003. Environmental Movements and Fisherfolk Participation on a Coastal Frontier, Palawan Island, Philippines. Athens (GA): University of Georgia (Ph.D. thesis).

Austin, R. and J.F. Eder, 2007. 'Environmentalism, Development, and Participation on Palawan Island, Philippines.' *Society and Natural Resources* 20: 363–371.

Bailey, C. and S. Jentoft, 1990. 'Hard Choices in Fisheries Development.' *Marine Policy* 14: 333–344.

Baum, G.A. and J.A. Maynard, 1976. 'Coron/Tagumpay: A Socio-Economic Study and Development Proposal.' Makati: South China Sea Fisheries Development and Coordinating Programme.

Bavinck, M., 2005. 'Understanding Fisheries Conflicts in the South—A Legal Pluralist Perspective.' *Society and Natural Resources* 18: 805–820.

Bell, S., K. Hampshire and S. Topalidou, 2007. 'The Political Culture of Poaching: A Case Study from Northern Greece.' *Biodiversity and Conservation* 16: 399–418.

Berkes, F., J. Colding and C. Folke (eds), 2003. *Navigating Social-Ecological Systems: Building Resilience for Complexity and Change.* Cambridge: Cambridge University Press.

BFAR (Bureau of Fisheries and Aquatic Resources), 2004. 'In Turbulent Seas: The Status of Philippine Marine Fisheries.' Cebu City: Bureau of Fisheries and Aquatic Resources and Department of Agriculture, Coastal Resource Management Project.

Billig, M.S., 2000. 'Institutions and Culture: Neo-Weberian Economic Anthropology.' *Journal of Economic Issues* 34: 771–788.

Blaikie, P., 1985. *The Political Economy of Soil Erosion in Developing Countries.* London: Longman.

Blaikie, P. and H. Brookfield, 1987. *Land Degradation and Society.* London: Methuen.

Blanc-Szanton, M.C., 1972. *A Right to Survive: Subsistence Marketing in a Lowland Philippine Town.* University Park (PA): Pennsylvania State University Press.

Borchgrevink, A., 2003. 'Ideas of Power in the Philippines.' *Cultural Dynamics* 15: 41–69.

Broad, R. and J. Cavanagh, 1993. *Plundering Paradise: The Struggle for the Environment in the Philippines.* Berkeley (CA): University of California Press.

Brockington, D., 2004. 'Community Conservation, Inequality and Injustice: Myths of Power in Protected Area Management.' *Conservation and Society* 2: 411–432.

Brookfield, H., L. Potter and Y. Byron, 1995. *In Place of the Forest: Environmental and Socio-Economic Transformation in Borneo and the Eastern Malay Peninsula.* New York: United Nations University Press.

Brosius, J.P., 1999. 'Analyses and Interventions: Anthropological Engagements with Environmentalism.' *Current Anthropology* 40: 277–309.

———, 2001. 'The Politics of Ethnographic Presence: Sites and Topologies in the Study of Transnational Environmental Movements.' In C. Crumley (ed.), *New Directions in Anthropology and Environment: Intersections.* Walnut Creek (CA): AltaMira Press.

———, 2006. 'What Counts as Local Knowledge in Global Environmental Assessments and Conventions?' In W.V. Reid, F. Berkes, T. Wilbanks and D. Capistrano (eds), *Bridging Scales and Epistemologies: Linking Local Knowledge and Global Science in Multi-Scale Assessments.* Washington (DC): Island Press.

Brosius, J.P., A.L. Tsing and C. Zerner (eds), 2005. *Communities and Conservation: Histories and Politics of Community-Based Natural Resource Management.* Walnut Creek (CA): AltaMira Press.

Brown, C. and M. Purcell, 2005. 'There's Nothing Inherent about Scale: Political Ecology, the Local Trap, and the Politics of Development in the Brazilian Amazon.' *Geoforum* 36: 607–624.

Bryant, R.L., 1992. 'Political Ecology: An Emerging Research Agenda in Third World Studies.' *Political Geography* 11: 12–36.

———, 2000. 'Politicized Moral Geographies: Debating Biodiversity Conservation and Ancestral Domain in the Philippines.' *Political Geography* 19: 673–705.

———, 2002. 'Non-Governmental Organizations and Governmentality: "Consuming" Biodiversity and Indigenous People in the Philippines.' *Political Studies* 50: 268–292.

————, 2005. *Nongovernmental Organizations in Environmental Struggles: Politics and the Making of Moral Capital in the Philippines.* New Haven (CT): Yale University Press.

Bulatao, J.C., 1964. 'Hiya.' *Philippine Studies* 12: 424–438.

Bulloch, H. and M. Fabinyi, 2009. 'Transnational Relationships, Transforming Selves: Filipinas Seeking Husbands Abroad.' *The Asia-Pacific Journal of Anthropology* 10: 129–142.

Butcher, J.G., 2004. *The Closing of the Frontier: A History of the Marine Fisheries of Southeast Asia c. 1850–2000.* Singapore: Institute of Southeast Asian Studies.

Calleja, N.C., 2009. 'Coron Islet to be Next Asian Tourism Jewel.' *Philippine Daily Inquirer*, 26 March.

Cannell, F., 1999. *Power and Intimacy in the Christian Philippines.* Cambridge: Cambridge University Press.

Carpenter, K.E. and V.G. Springer, 2005. 'The Center of the Center of Marine Shore Fish Biodiversity: The Philippine Islands.' *Environmental Biology of Fishes* 72: 467–480.

Carrier, J.G., 2001. 'Limits of Environmental Understanding: Action and Constraint.' *Journal of Political Ecology* 8: 25–44.

Cervino, J.M., R.L. Hayes, M. Honovich, T.J. Goreau, S. Jones and P.J. Rubec, 2003. 'Changes in Zooxanthellae Density, Morphology, and Mitotic Index in Hermatypic Corals and Anemones Exposed to Cyanide.' *Marine Pollution Bulletin* 46: 573–586.

Christie, P., 2005a. 'Is Integrated Coastal Management Sustainable?' *Ocean & Coastal Management* 48: 208–232.

————, 2005b. 'Observed and Perceived Environmental Impacts of Marine Protected Areas in Two Southeast Asia Sites.' *Ocean & Coastal Management* 48: 252–270.

Christie, P., D.L. Fluharty, A.T. White, L. Eisma-Osorio and W. Jatulan, 2007. 'Assessing the Feasibility of Ecosystem-Based Fisheries Management in Tropical Contexts.' *Marine Policy* 31: 239–250.

Christie, P., K. Lowry, A.T. White, E.G. Oracion, L. Sievanen, R.S. Pomeroy, R.B. Pollnac, J.M. Patlis and R-L.V. Eisma, 2005. 'Key Findings from a Multidisciplinary Examination of Integrated Coastal Management Process Sustainability.' *Ocean & Coastal Management* 48: 468–483.

Christie, P., R.B. Pollnac, D.L. Fluharty, M.A. Hixon, G.K. Lowry, R. Mahon, D. Pietri, B.N. Tissot, A.T. White, N. Armada, and R. Eisma-Osorio, 2009. 'Tropical Marine EBM Feasibility: A Synthesis of Case Studies and Comparative Analyses.' *Coastal Management* 37: 374–385.

Christie, P., A.T. White and E. Deguit, 2002. 'Starting Point or Solution? Community-Based Marine Protected Areas in the Philippines.' *Journal of Environmental Management* 66: 441–454.

Christie, P., A.T. White, B. Stockwell and R.C. Jadloc, 2003. 'Links between Environmental Conditions and Integrated Coastal Management Sustainability.' *Silliman Journal* 44: 285–323.

CI (Conservation International), 2003. 'Analysis of the Benefits and Costs of the Live Reef Food Fish Trade in the Calamianes, Palawan, Philippines.' Quezon City: Conservation International–Philippines.

Clarke, G. and M. Sison, 2003. 'Voices from the Top of the Pile: Elite Perceptions of Poverty and the Poor in the Philippines.' *Development and Change* 34: 215–242.

Clifford, J. and G.E. Marcus, 1986. *Writing Culture: The Politics and Poetics of Ethnography*. Berkeley (CA): University of California Press.

CLOA (Calamianes Live-fish Operators Association), 2006. 'The Calamianes Live Fish Industry.' Coron: Unpublished report.

Comim, F., 2008. 'Poverty and Environment Indicators.' Cambridge: Capability and Sustainability Centre (report for UNDP-UNEP Poverty and Environment Initiative).

Courtney, C. and A.T. White, 2000. 'Integrated Coastal Management in the Philippines: Testing New Paradigms.' *Coastal Management* 28: 39–53.

Cruz, R., 2005. 'Baseline Assessment of the Capture Fisheries in the FISH Project's Focal Areas: Coron Bay.' Cebu City: FISH Project.

Dalabajan, D., 2000. 'The Healing of a Tagbanua Ancestral Homeland.' In E.M. Ferrer, L.P. de la Cruz and G.R. Newkirk (eds), *Hope Takes Root: Community-Based Coastal Resource Management Stories from Southeast Asia*. Quezon City: Community Based Coastal Resources Management Resource Center and Coastal Resources Research Network.

———, 2005. 'Fixing the Broken Net: Improving Enforcement of Laws Regulating Cyanide Fishing in the Calamianes Group of Islands, Philippines.' *SPC Live Reef Fish Information Bulletin* 15: 3–12.

————, 2009. 'Of Crimes and No Punishments: Fisheries Law Offences and the Criminal Justice System in Calamianes Group of Islands in the Province of Palawan.' In R.D. Anda and D.A. Dalabajan (eds), *Against the Tide: Enforcement and Governance in the Sulu Sulawesi Seas*. Puerto Princesa City: Futuristic Printing Press.

Dannhaeuser, N., 1983. *Contemporary Trade Strategies in the Philippines*. New Brunswick (NJ): Rutgers University Press.

Davis, W.G., 1973. *Social Relations in a Philippine Market: Self-Interest and Subjectivity*. Berkeley (CA): University of California Press.

Depondt, F. and E. Green, 2006. 'Diving User Fees and the Financial Sustainability of Marine Protected Areas: Opportunities and Impediments.' *Ocean & Coastal Management* 49: 188–202.

Descola, P. and G. Pálsson, 1996. *Nature and Society: Anthropological Perspectives*. London: Routledge.

Doronila, A., 2001. *The Fall of Joseph Estrada: The Inside Story*. Manila: Anvil Publishing and Philippine Daily Inquirer.

Dressler, W.H., 2009. *Old Thoughts in New Ideas: State Conservation Measures, Livelihood and Development on Palawan Island, The Philippines*. Quezon City: Ateneo de Manila University Press.

Dressler, W.H. and M. Fabinyi, 2011. 'Farmer Gone Fish'n? Swidden Decline and the Rise of Grouper Fishing on Palawan, Philippines.' *Journal of Agrarian Change* 11: 536–555.

Dumont, J-P., 1992. *Visayan Vignettes: Ethnographic Traces of a Philippine Island*. Chicago: University of Chicago Press.

————, 1993. 'The Visayan Male Barkada: Manly Behaviour and Male Identity on a Philippine Island.' *Philippine Studies* 41: 401–436.

————, 1995. 'Far from Manila: Political Identities on a Philippine Island.' *Anthropological Quarterly* 68: 14–20.

Dwyer, K., 1987. *Moroccan Dialogues: Anthropology in Question*. Illinois: Waveland Press.

Eder, J.F., 1999. *A Generation Later: Household Strategies and Economic Change in the Rural Philippines*. Quezon City: Ateneo de Manila University Press.

————, 2003. 'Of Fishers and Farmers: Ethnicity and Resource Use in Coastal Palawan.' *Philippine Quarterly of Culture & Society* 31: 207–225.

————, 2004. 'Who are the Cuyonon? Ethnic Identity in the Modern Philippines.' *Journal of Asian Studies* 63: 625–647.

————, 2005. 'Coastal Resource Management and Social Differences in Philippine Fishing Communities.' *Human Ecology* 33: 147–169.

————, 2006. 'Gender Relations and Household Economic Planning in the Rural Philippines.' *Journal of Southeast Asian Studies* 37: 397–413.

————, 2008. *Migrants to the Coasts: Livelihood, Resource Management, and Global Change in the Philippines.* Wadsworth (CT): Cengage Learning.

Eder, J.F. and J.O. Fernandez (eds), 1996. *Palawan at the Crossroads: Development and the Environment on a Philippine Frontier.* Quezon City: Ateneo de Manila University Press.

ELAC (Environmental Legal Assistance Center), 2004. *Mending Nets: A Handbook on the Prosecution of Fishery and Coastal Law Violations.* Cebu City: Environmental Legal Assistance Center.

Escobar, A., 1995. *Encountering Development: The Making and Unmaking of the Third World.* Princeton (NJ): Princeton University Press.

————, 1999. 'After Nature: Steps to an Anti-essentialist Political Ecology.' *Current Anthropology* 40: 1–30.

Esguerra, E.M., 1999. Forests, People and Palawan: The Challenges of Implementing the *Strategic Environmental Plan (SEP) Law.* Canberra: Australian National University (Ph.D. thesis).

Fabinyi, M., 2010. 'The Intensification of Fishing and the Rise of Tourism: Competing Coastal Livelihoods in the Calamianes Islands, Philippines.' *Human Ecology* 38: 415–427.

————, forthcoming. 'Fishing and Socio-Economic Change in the Calamianes Islands.' In J.F. Eder and O. Evangelista (eds), *Palawan and its Global Networks.*

Fabinyi, M. and D. Dalabajan, 2011. 'Policy and Practice in the Live Reef Fish for Food Trade: A Case Study from Palawan, Philippines.' *Marine Policy* 35: 37–38.

Fabinyi, M., M. Knudsen and S. Segi, 2010. 'Social Complexity, Ethnography and Coastal Resource Management in the Philippines.' *Coastal Management* 38: 617–632.

Fegan, B., 1982. 'The Social History of a Central Luzon Barrio.' In A.W. McCoy and E.C. de Jesus (eds), *Philippine Social History: Global Trade and Local Transformations*. Quezon City: Ateneo de Manila University Press.

FISHBASE, 2008. '*Siganus argenteus*.' Viewed 15 January 2009 at http://www.fishbase.org/Summary/speciesSummary.P?ID=4614&genusname=Siganus&speciesname=argenteus

Filer, C., 2004. 'The Knowledge of Indigenous Desire: Disintegrating Conservation and Development in Papua New Guinea.' In A. Bicker, P. Sillitoe and J. Pottier (eds), *Development and Local Knowledge: New Approaches to Issues in Natural Resource Management, Conservation and Agriculture*. London: Routledge.

————, 2009. 'A Bridge Too Far: The Knowledge Problem in the Millennium Assessment.' In J. Carrier and P. West (eds), *Virtualism, Governance and Practice: Vision and Execution in Environmental Conservation*. New York: Berghahn Books.

Firth, R., 1966. *Malay Fishermen: Their Peasant Economy*. London: Routledge & Kegan Paul (2nd edition).

FISH (Fisheries Improved for Sustainable Harvest) Project, 2005. 'Consolidated Report: Baseline Assessment of the Capture Fisheries and Marine Protected Areas (Reef Habitats) in the FISH Project's Focal Areas: Coron Bay, Danajon Bank, Lanuza Bay and Tawi-Tawi Bay.' Cebu City: Philippines Department of Agriculture.

Flyvbjerg, B., 2001. *Making Social Science Matter: Why Social Inquiry Fails and How It Can Succeed Again*. Cambridge: Cambridge University Press.

Foale, S., 1998. 'Assessment and Management of the Trochus Fishery at West Nggela, Solomon Islands: An Interdisciplinary Approach.' *Ocean & Coastal Management* 40: 187–205.

Fox, R.B., 1982. *Religion and Society among the Tagbanua of Palawan Island, Philippines*. Manila: Philippines National Museum.

Friedmann, J.R.P., 1966. *Regional Development Policy: A Case Study of Venezuela*. Cambridge (MA): MIT Press.

Galvez, R.E., T.G. Hingco, C. Bautista and M.T. Tungpalan, 1989. 'Sociocultural Dynamics of Blast Fishing and Sodium Cyanide Fishing in Two Fishing Villages in the Lingayen Gulf Area.' In G.T. Silvestre, E. Miclat and C. Thia-Eng (eds), *Towards Sustainable Development of the Coastal Resources of Lingayen Gulf, Philippines*. Makati: International Center for Living Aquatic Resources Management.

Geertz, C., 1973. *The Interpretation of Cultures*. London: Basic Books.

GMANews TV, 2007. '3 Sacks of Ammonium Nitrate Seized in Manila.' Viewed 4 August 2007 at http://www.gmanews.tv/story/53941/3-sacks-of-ammonium-nitrate-seized-in-Manila#

Gov. Ph. News, 2008. 'PGMA opens Busuanga's "Holiday Airport".' Viewed 18 November 2008 at http://www.gov.ph/news/?i=22830

Gray, L.C. and W.G. Moseley, 2005. 'A Geographical Perspective on Poverty-Environment Interactions.' *The Geographical Journal* 171: 9–23.

Green, S.J., 2004. 'Establishing a Mechanism for the Collection of User Fees, Administration and Management of a Marine Parks in the Municipality of Coron.' Palawan: Sustainable Environmental Management Program, Northern Palawan.

Green, S.J., A.T. White, J.O. Flores, M.F. Carreon III and A.E. Sia, 2003. *Philippine Fisheries in Crisis: A Framework for Management*. Cebu City: Philippines Department of Environment and Natural Resources, Coastal Resource Management Project.

Hall, S., 1997. 'The Work of Representation.' In S. Hall (ed.), *Representation: Cultural Representations and Signifying Practices*. London: Sage.

Hampshire, K., S. Bell, G. Wallace and F. Stepukonis, 2004. '"Real" Poachers and Predators: Shades of Meaning in Local Understandings of Threats to Fisheries.' *Society and Natural Resources* 17: 305–318.

Hendriks, M., 1994. 'Trade Arrangements and Interlinked Credit in the Philippines.' In F.J.A. Bouman and O. Hospes (eds), *Financial Landscapes Reconstructed: The Fine Art of Mapping Development*. Boulder (CO): Westview Press.

Herre, A.W.C.T., 1948. 'Outlook for Philippine Fisheries.' *Far Eastern Survey* 17: 277–278.

Hilborn, R., K. Stokes, J-J. Maguire, T. Smith, L.W. Botsford, M. Mangel, J. Orensanz, A. Parma, J. Rice, J. Bell, K.L. Cochrane, S. Garcia, S.J. Hall, G.P. Kirkwood, K. Sainsbury, G. Stefansson and C. Walters, 2004. 'When Can Marine Reserves Improve Fisheries Management?' *Ocean & Coastal Management* 47: 197–205.

Hirtz, F., 2003. 'It Takes Modern Means to be Traditional: On Recognizing Indigenous Cultural Communities in the Philippines.' *Development and Change* 34: 887–914.

Hollnsteiner, M., 1963. *The Dynamics of Power in a Philippine Municipality*. Quezon City: University of the Philippines.

———, 1970. 'Reciprocity in the Lowland Philippines.' In F. Lynch and A. de Guzman II (eds), op. cit.

Hviding, E., 1996. *Guardians of Marovo Lagoon: Practice, Place and Politics in Maritime Melanesia*. Honolulu: University of Hawai'i Press.

Hyndman, D., 1994. 'A Sacred Mountain of Gold: The Creation of a Mining Resource Frontier in Papua New Guinea.' *Journal of Pacific History* 29: 203–222.

Ileto, R.C., 1979. *Pasyon and Revolution: Popular Movements in the Philippines, 1840–1910*. Quezon City: Ateneo de Manila University Press.

Ingles, J. 2000. 'Fisheries of the Calamianes Islands, Palawan Province, Philippines.' In T.B. Werner and G.R. Allen (eds), op. cit.

Ingold, T., 1993. 'Globes and Spheres: The Topology of Environmentalism.' In K. Milton (ed.), *Environmentalism: The View from Anthropology*. London: Routledge.

IPC (Institute of Philippine Culture), 2005. *The Vote of the Poor: Modernity and Tradition in People's Views of Leadership and Elections*. Quezon City: Institute of Philippine Culture.

Jimenez-David, R., 2007. 'China's Role in the Hoi Wan Incident.' *Philippine Daily Inquirer*, 1 January.

Jocano, F.L., 1975. *Slum as a Way of Life: A Study of Coping Behaviour in an Urban Environment*. Quezon City: University of the Philippines Press.

Johnson, D.S., 2006. 'Category, Narrative and Value in the Governance of Small-Scale Fisheries.' *Marine Policy* 30: 747–756.

Junker, L.L., 1999. *Raiding, Trading, and Feasting: The Political Economy of Philippine Chiefdoms*. Honolulu: University of Hawai'i Press.

Kasuya, Y., and N.G. Quimpo (eds), 2010. *The Politics of Change in the Philippines*. Pasig City: Anvil Press.

Kaut, C., 1961. 'Utang Na Loob: A System of Contractual Obligation among Tagalogs.' *Southwestern Journal of Anthropology* 17: 256–273.

Kelly, P.F., 2000. *Landscapes of Globalisation: Human Geographies of Economic Change in the Philippines*. London: Routledge.

Kerkvliet, B.J., 1990. *Everyday Politics in the Philippines: Class and Status Relations in a Central Luzon Village.* Berkeley (CA): University of California Press.

———, 1995. 'Toward a More Comprehensive Analysis of Philippine Politics: Beyond the Patron-Client, Factional Framework'. *Journal of Southeast Asian Studies* 26: 401–419.

———, 1996. 'Contested Meanings of Elections in the Philippines.' In R.H. Taylor (ed.), *The Politics of Elections in Southeast Asia.* Cambridge: Cambridge University Press.

Kerkvliet, B.J. and R.B. Mojares (eds), 1991. *From Marcos to Aquino: Local Perspectives on Political Transition in the Philippines.* Quezon City: Ateneo de Manila University Press.

Knudsen, M. 2008. 'Appropriation of Coastal Space and Community Formation on a Philippine Island.' Paper presented at the Association of Social Anthropologists of the UK and Commonwealth Conference, Auckland, 8–12 December.

Kurlansky, M., 1999. *Cod: A Biography of the Fish that Changed the World.* London: Vintage.

Lahiri-Dutt, K., 2003. 'Informal Coal Mining in Eastern India: Evidence from the Raniganj Coalbelt.' *Natural Resources Forum* 27: 68–77.

Landé , C., 1965. *Leaders, Factions and Parties: The Structure of Philippine Politics.* New Haven (CT): Yale University Press.

Lawrence, K., 2002. Negotiated Biodiversity Conservation for Local Social Change: A Case Study of Northern Palawan, Philippines. London: London University, Kings College (Ph.D. thesis).

Lebel, L., J.M. Anderies, B. Campbell, C. Folke, S. Hatfield-Dodds, T.P. Hughes and J. Wilson, 2006. 'Governance and the Capacity to Manage Resilience in Regional Social-Ecological Systems.' *Ecology and Society* 11(1): 19. Viewed 15 November 2011 at http://www.ecologyandsociety.org/vol11/iss1/art19/

Lewis, M.W., 1992. *Wagering the Land: Ritual, Capital, and Environmental Degradation in the Cordillera of Northern Luzon.* Berkeley (CA): University of California Press.

Li, T.M., 2005. 'Beyond "the State" and Failed Schemes.' *American Anthropologist* 107: 383–394.

————, 2007. *The Will to Improve: Governmentality, Development, and the Practice of Politics*. Durham (NC): Duke University Press.

Lowe, C., 2000. 'Global Markets, Local Injustice in Southeast Asian Seas: The Live Fish Trade and Local Fishers in the Togean Islands.' In C. Zerner (ed.), *People, Plants, and Justice: The Politics of Nature Conservation*. New York: Columbia University Press.

————, 2006. *Wild Profusion: Biodiversity Conservation in an Indonesian Archipelago*. Princeton (NJ): Princeton University Press.

Lynch, F., 1970. 'Social Acceptance Reconsidered.' In F. Lynch and A. de Guzman II (eds), op.cit.

Lynch, F. and A. de Guzman II (eds), 1970. *Four Readings on Filipino Values*. Quezon City: Ateneo de Manila University Press.

Macintyre, M., 2008. 'Police and Thieves, Gunmen and Drunks: Problems with Men and Problems with Society in Papua New Guinea.' *The Australian Journal of Anthropology* 19: 179–193.

Majanen, T., 2007. 'Resource Use Conflicts in Mabini and Tingloy, the Philippines.' *Marine Policy* 31: 480–487.

Manila Times, 2006. 'Fishers Decry Live Fish Ruling.' *Manila Times*, 20 December.

Marcus, G.E., 1995. 'Ethnography in/of the World System: The Emergence of Multi-Sited Ethnography.' *Annual Review of Anthropology* 24: 95–117.

Mayo-Anda, G., D. Dalabajan and N.C. Lasmarias, 2003. 'Deterrent Value of Law Enforcement on Dynamite and Cyanide Fishing: An Enforcement Economics Study of the Calamianes Group of Islands, Palawan, Philippines.' Unpublished report for Conservation International.

McCarthy, J.W., 2006. *The Fourth Circle: A Political Ecology of Sumatra's Rainforest Frontier*. Stanford (CA): Stanford University Press.

McCoy, A.W., 1999. *Closer Than Brothers: Manhood at the Philippine Military Academy*. Manila: Anvil Publishing.

McDermott, M.H., 2000. Boundaries and Pathways: Indigenous Identity, Ancestral Domain, and Forest Use in Palawan, the Philippines. Berkeley (CA): University of California (Ph.D. thesis).

McKay, D., 2003. 'Cultivating New Local Futures: Remittance Economies and Land-Use Patterns in Ifugao, Philippines.' *Journal of Southeast Asian Studies* 34: 285–306.

McShane, T.O., P.D. Hirsch, T.C. Trung, A.N. Songorwa, A. Kinzig, B. Monteferri, D. Mutekanga, H.V. Thang, J.L. Dammert, M. Pulgar-Vidal, M. Welch-Devine, J.P. Brosius, P. Coppolillo and S.O'Connor, 2011. 'Hard Choices: Making Trade-Offs between Biodiversity Conservation and Human Well-Being.' *Biological Conservation* 144: 966–972.

McShane, T.O. and M.P. Wells (eds), 2004. *Getting Biodiversity Projects to Work: Towards More Effective Conservation and Development*. New York: Columbia University Press.

MEA (Millennium Ecosystem Assessment), 2003. *Ecosystems and Human Well-Being: A Framework for Assessment*. Washington (DC): Island Press.

Mintz, S.W., 1961. 'Pratik: Haitian Personal Economic Relationships.' In *Proceedings of the 1961 Annual Spring Meeting of the American Ethnological Society*. Seattle: University of Washington Press.

Moore, D.S., 1996. 'Marxism, Culture, and Political Ecology.' In R. Peet and M. Watts (eds), op. cit.

Moseley, W.G., 2005. 'Global Cotton and Local Environmental Management: The Political Ecology of Rich and Poor Small-Hold Farmers in Southern Mali.' *Geographical Journal* 171: 36–55.

Mulder, N., 1997. *Inside Philippine Society: Interpretations of Everyday Life*. Quezon City: New Day Publishers.

———, 2000. *Filipino Images: Culture of the Public World*. Quezon City: New Day Publishers.

Municipality of Coron, 2002. 'History of Coron'. In *Paraiso ng Kalikasan, Hiyas ng Palawan* [*Paradise of Nature, Jewel of Palawan*]. Coron: Municipality of Coron.

NSO (National Statistics Office), 2010. '2007 Census Data.' Viewed 3 December 2010 at http://www.census.gov.ph/

Novellino, D., 2003. 'Contrasting Landscapes, Conflicting Ontologies: Assessing Environmental Conservation on Palawan Island (the Philippines).' In D.G. Anderson and E. Berglund (eds), *Ethnographies of Conservation: Environmentalism and the Distribution of Privilege*. London: Berghahn Books.

————, 2007. '"Talking About Kultura and Signing Contracts": The Bureaucratization of the Environment on Palawan Island (the Philippines).' In C.A. Maida (ed.), *Sustainability and Communities of Place*. London: Berghahn Books.

Novellino, D., and W.H. Dressler, 2010. 'The Role of "Hybrid" NGOs in the Conservation and Development of Palawan Island, the Philippines.' *Society and Natural Resources* 2: 165–180.

Olive, S., 1996. 'The Usefulness and Limitations of Common Property Concepts for Understanding Coastal Resource Management in the Philippines.' Paper presented at the 5th International Philippine Studies Conference, Honolulu, 14–16 April.

Oracion, E.G., 2003. 'The Dynamics of Stakeholder Participation in Marine Protected Area Development: A Case Study in Batangas, Philippines.' *Silliman Journal* 44: 95–137.

————, 2006. Appropriating Space: An Ethnography of Local Politics and Cultural Politics of Marine Protected Areas in Dauin, Negros Oriental, Philippines. Cebu City: University of San Carlos (Ph.D. thesis).

Oracion, E.G., M.L. Miller and P. Christie, 2005. 'Marine Protected Areas for Whom? Fisheries, Tourism, and Solidarity in a Philippine Community.' *Ocean & Coastal Management* 48: 393–410.

Ortner, S., 1995. 'Resistance and the Problem of Ethnographic Refusal.' *Comparative Studies in Society and History* 37: 173–193.

Padilla, J.E., S. Mamauag, G. Braganza, N. Brucal, D. Yu and A. Morales, 2003. 'Sustainability Assessment of the Live Reef-Fish for Food Industry in Palawan Philippines.' Quezon City: WWF–Philippines.

PAFID (Philippine Association for Intercultural Development), 2000. 'Mapping the Ancestral Lands and Waters of the Calamian Tagbanwa of Coron, Northern Palawan.' In Bennagen, P.L. and Royo, A.G. (eds), *Mapping the Earth, Mapping Life*. Quezon City: Legal Rights and Natural Resources Center.

Palawan Department of Tourism, 2006. 'Tourist Arrivals.' Puerto Princesa City: Unpublished report.

————, 2007. 'The Official Tourism Website of the Province of Palawan.' Viewed 15 May 2007 at http://www.palawan.tourism.com

Palawan Provincial Government, 2000. 'Human Development Report.' Puerto Princesa City: Provincial Planning and Development Office.

Palawan Times, 2009. 'Coron Airport in Northern Palawan Attracts More Tourists.' Viewed 13 October 2009 at: http:// thepalawantimes.wordpress. com/2009/08/04/coron-airport-in- northern-palawan-attracts-more-tourists/

Pálsson, G., 1991. *Coastal Economies, Cultural Accounts: Human Ecology and Icelandic Discourse*. Manchester and New York: Manchester University Press.

Pálsson, G. and E.P. Durrenberger, 1990. 'Systems of Production and Social Discourse: The Skipper Effect Revisited.' *American Anthropologist* 92: 130–141.

Paulson, S. and L. Gezon (eds), 2005. *Political Ecology across Spaces, Scales, and Social Groups*. New Brunswick: Rutgers University Press.

Paulson, S., L. Gezon and M. Watts, 2005. 'Politics, Ecologies, Genealogies.' In S. Paulson and L. Gezon (eds), op. cit.

Peet, R. and M. Watts (eds), 1996. *Liberation Ecologies: Environment, Development, Social Movements*. London: Routledge.

Peluso, N.L., 1992. *Rich Forests, Poor People: Resource Control and Resistance in Java*. Berkeley: University of California Press.

Pertierra, R., 2002. *The Work of Culture*. Manila: De La Salle University Press.

Pinches, M., 1991. 'The Working-Class Experience of Shame, Inequality and People Power in Tatalon, Manila.' In B.J. Kerkvliet and R.B. Mojares (eds), op. cit.

———, 2008. 'Residence, Intimacy and Resilience: Urban Poor Housing and Kinship in Tatalon 1953–2007.' Paper presented at the Association of Social Anthropologists of the UK and Commonwealth Conference, Auckland, 8–12 December.

Plattner, S., 1983. 'Economic Custom in a Competitive Marketplace.' *American Anthropologist* 85: 848–858.

Pollnac, R.B, B.R. Crawford and M.L.G Gorospe, 2001 (Pollnac et al. 2001a). 'Discovering Factors that Influence the Success of Community-Based Marine Protected Areas in the Visayas, Philippines.' *Ocean & Coastal Management* 44: 683–710.

Pollnac, R.B. and J.C. Johnson, 2005. 'Folk Management and Conservation of Marine Resources.' In N. Kishigami and J.M. Savelle (eds), *Indigenous Use and Management of Marine Resources*. Osaka: National Museum of Ethnology.

Pollnac, R.B., R.S. Pomeroy and I.H.T. Harkes, 2001 (Pollnac et al. 2001b). 'Fishery Policy and Job Satisfaction in Three Southeast Asian Fisheries.' *Ocean & Coastal Management* 44: 531–44.

Polo, J.B., 1985. 'Of Metaphors and Men: The Binlayan Fishcorral Ritual as a Contract in a Social Spectrum.' *Philippine Sociological Review* 33: 54–63.

Pomeroy, R.S., M.D. Pido, J. Pontillas, B.S. Francisco, A.T. White and G.T. Silvestre, 2005. 'Evaluation of Policy Options for the Live Reef Food Fish Trade: Focus on Calamianes Islands and Palawan Province, Philippines, with Implications for National Policy.' Palawan: Palawan Council for Sustainable Development, Fisheries Improved for Sustainable Harvest Project, and Provincial Government of Palawan.

Pomeroy, R.S., M.D. Pido, J. Pontillas, B.S. Francisco, A.T. White, E.M.C. Ponce de Leon and G.T. Silvestre, 2008. 'Evaluation of Policy Options for the Live Reef Food Fish Trade in the Province of Palawan, Western Philippines.' *Marine Policy* 32: 55–65.

Potter, L.M., 1997. 'A Forest Product Out of Control: *Gutta percha* in Indonesia and the Wider Malay World, 1845–1915'. In P. Boomgaard, F. Colombin and D. Henley (eds), *Paper Landscapes: Explorations in the Environmental History of Indonesia*. Leiden: KITLV Press.

Rafael, V.L., 1988. *Contracting Colonialism: Translation and Christian Conversion in Tagalog Society under Early Spanish Rule*. Quezon City: Ateneo de Manila University Press.

———, 2000. *White Love and Other Events in Filipino History*. Quezon City: Ateneo de Manila University Press.

———, 2010. 'Welcoming What Comes: Sovereignty and Revolution in the Colonial Philippines.' *Comparative Studies in Society and History* 52: 157–179.

Ratner, B.D., 2006. 'Community Management by Decree? Lessons from Cambodia's Fisheries Reform.' *Society and Natural Resources* 19: 79–86.

Reed, D. and P. Tharakan, 2004. 'Developing and Applying Poverty Environment Indicators.' Washington (DC): WWF Macroeconomics Program Office.

Rigg, J., 2006. 'Land, Farming, Livelihoods, and Poverty: Rethinking the Links in the Rural South.' *World Development* 34(1): 180–202.

Robben, A., 1989. *Sons of the Sea Goddess: Economic Practice and Discursive Conflict in Brazil*. New York: Columbia University Press.

Robbins, P., 2004. *Political Ecology: A Critical Introduction*. Cornwall: Blackwell Publishing.

Roberts, C., 2007. *An Unnatural History of the Sea*. Washington (DC): Island Press.

Roberts, C. and J. Hawkins, 2003. 'Fully Protected Marine Reserves: A Guide.' Washington (DC): WWF Endangered Seas Campaign.

Robinson, J. 2006 [1962]. *Economic Philosophy*. Chicago: Aldine Publishing Company.

Roces, M., 2001. *Kinship Politics in Post-War Philippines: The Lopez Family, 1946–2000*. Manila: De La Salle University Press.

Russ, G.R., 2002. 'Yet Another Review of Marine Reserves as Reef Fishery Management Tools.' In P.F. Sale (ed.), op.cit.

Russell, B., 2004 [1946]. *History of Western Philosophy*. London: Routledge.

Russell, S.D., 1987. 'Middlemen and Moneylending: Relations of Exchange in a Highland Philippine Economy.' *Journal of Anthropological Research* 42: 139–161.

———, 1997. 'Class Identity, Leadership Style, and Political Culture in a Tagalog Coastal Community.' *Pilipinas* 28: 79–95.

Russell, S.D., and R.T. Alexander, 1996. 'The Skipper Effect Debate: Views from a Philippine Fishery.' *Journal of Anthropological Research* 52: 433–459.

———, 2000. 'Of Beggars and Thieves: Customary Sharing of the Catch and Informal Sanctions in a Philippine Fishery.' In E.P. Durrenberger and T.D. King (eds), *State and Community in Fisheries Management: Power, Policy and Practice*. Westport (CT): Bergin & Garvey.

Rutten, R., 1991. 'Intepreting Credit Relations: Do Philippine Artisans Enjoy Credit or Suffer Indebtedness?' *Netherlands Journal of Social Science* 27: 108–117.

Sabetian, A. and S. Foale, 2006. 'Evolution of the Artisanal Fisher: Case Studies from Solomon Islands and Papua New Guinea.' *SPC Traditional Marine Resource Management and Knowledge Information Bulletin* 20: 3–10.

Sahlins, M., 1972. *Stone Age Economics*. New York: Aldine de Gruyter.

———, 1985. *Islands of History*. Chicago: University of Chicago Press.

Sale, P.F., 2002. 'The Science We Need to Develop for More Effective Management.' In P.F. Sale (ed.), op.cit.

———— (ed.), 2002. *Coral Reef Fishes: Dynamics and Diversity in a Complex Ecosystem*. Oxford and St Louis: Academic Press.

Scales, H., A. Balmford, M. Liu, Y. Sadovy, and A. Manica, 2006. 'Keeping Bandits at Bay?' *Science* 313: 612–613.

Scott, J.C., 1976. *The Moral Economy of the Peasant: Rebellion and Subsistence in South-East Asia*. New Haven: Yale University Press.

————, 1985. *Weapons of the Weak: Everyday Forms of Peasant Resistance*. New Haven (CT): Yale University Press.

————, 1998. *Seeing like a State: How Certain Schemes to Improve the Human Condition have Failed*. New Haven (CT): Yale University Press.

Scott, J.C. and B.J. Kerkvliet, 1977. 'How Traditional Rural Patrons Lose Legitimacy: A Theory with Special Reference to Southeast Asia.' In S.W. Schmidt (ed.), *Friends Followers and Factions: A Reader in Political Clientielism*. Berkeley (CA): University of California Press.

Scott, W.H., 1982. *Cracks in the Parchment Curtain and Other Essays in Philippine History*. Quezon City: New Day Publishers.

————, 1983. 'Oripun and Alipun in the Philippines.' In A. Reid (ed.), *Slavery, Bondage and Dependency in Southeast Asia*. St Lucia: University of Queensland Press.

Seki, K., 2004. 'Maritime Migration in the Visayas: A Case Study of the Dalaguetenon Fisherfolk in Cebu.' In H. Umehara and G.M. Bautista (eds), *Communities at the Margins: Reflections on Social, Economic and Environmental Change in the Philippines*. Quezon City: Ateneo de Manila University Press.

————, 2009. 'Green Neoliberalism, Ecogovernmentality, and Emergent Community: A Case of Coastal Resource Management in Palawan, the Philippines.' *Philippine Studies* 57: 901–936.

Siar, S.V., 2003. 'Knowledge, Gender, and Resources in Small-Scale Fishing: The Case of Honda Bay, Palawan, Philippines.' *Environmental Management* 31: 569–80.

Spoehr, A., 1980. *Protein from the Sea: Technological Change in Philippine Capture Fisheries*. Pittsburgh (PA): University of Pittsburgh.

Soulé, M.E. and J. Terborgh, 1999. 'The Policy and Science of Regional Conservation.' In M.E. Soulé and J. Terborgh (eds), *Continental Conservation: Scientific Foundations of Regional Reserve Networks*. Washington (DC): Island Press.

Stacey, N., 2007. *Boats to Burn: Bajo Fishing Activity in the Australian Fishing Zone*. Canberra: ANU E Press (Asia-Pacific Environment Monograph 2).

Steward J.H., 1955. *Theory of Culture Change: The Methodology of Multilinear Evolution*. Urbana: University of Illinois Press.

Swift, O., 2006. 'Women in a Watery World? Flexibilities of Filipino International Seafaring.' Paper presented at the 28th Annual Conference of the Anthropological Association of the Philippines, Dumaguete, 26–28 October.

Szanton, D.L., 1971. *Estancia in Transition: Economic Growth in a Rural Philippine Community*. Quezon City: Ateneo de Manila Press.

Tsing, A.L., 2005. *Friction: An Ethnography of Global Connection*. Princeton (NJ): Princeton University Press.

Turner, F.J., 1996 [1920]. *The Frontier in American History*. New York: Courier Dover Publications.

Turner, V., 1975. 'Symbolic Studies.' *Annual Review of Anthropology* 4: 145–161.

Turton, A., 1986. 'Patrolling the Middle-Ground: Methodological Perspectives on "Everyday Peasant Resistance".' *Journal of Peasant Studies* 13(2): 36–48.

Tyler, S.A., 1986. 'Post-Modern Ethnography: From Document of the Occult to Occult Document.' In Clifford, J. and G.E. Marcus (eds), op. cit.

Ushijima, I. and C. Neri Zayas (eds), 1994. *Fishers of the Visayas: Visayas Maritime Anthropological Studies 1: 1991–1993*. Quezon City: CSSP Publications and University of the Philippines Press.

Van Helden, F., 2004. '"Making Do": Integrating Ecological and Societal Considerations for Marine Conservation in a Situation of Indigenous Resource Tenure.' In L.E. Visser (ed.), *Challenging Coasts: Transdisciplinary Excursions into Integrated Coastal Zone Development*. Amsterdam: Amsterdam University Press.

Vayda, A.P., 1969. 'Expansion and Warfare among Swidden Agriculturalists.' In A.P. Vayda (ed.), *Environment and Cultural Behavior*. New York: Natural History Press.

Veloro, C.E., 1994. 'Suwerte and Diskarte: Notions of Fishing, Success and Social Relations in Two Palawan Villages.' In I. Ushijima and C. Neri Zayas (eds), op. cit.

Villanueva Jr, A., 2008. 'Coron, San Vicente Towns Emerge as Palawan's Tourism Growth Centers.' *Palawan Times*, 18 May.

Walley, C.J., 2004. *Rough Waters: Nature and Development in an East African Marine Park*. Princeton (NJ): Princeton University Press.

WCED (World Commission on Environment and Development), 1987. *Our Common Future*. Oxford: Oxford University Press.

Weeks, R., G.R. Russ, A.C. Alcala, A.T. White, 2010. 'Effectiveness of Marine Protected Areas in the Philippines for Biodiversity Conservation.' *Conservation Biology* 24: 531–540.

Weeratunge, N., 2010. 'Gleaner, Fisher, Trader, Processor: Understanding Gendered Employment in Fisheries and Aquaculture.' *Fish and Fisheries* 11: 405–420.

Wells, M.P., T.O. McShane, H.T. Dublin, S. O'Connor and K.H. Redford, 2004. 'The Future of Integrated Conservation and Development Projects: Building on What Works.' In T.O. McShane and M.P. Wells (eds), op.cit.

Werner, T.B. and G.R. Allen, 2000. 'A Rapid Marine Biodiversity Assessment of the Calamianes Islands, Palawan Province, Philippines.' Washington (DC): Conservation International.

West, P., 2005. 'Translation, Value and Space: Theorizing an Ethnographic and Engaged Environmental Anthropology.' *American Anthropologist* 107: 632–642.

White, A.T., P. Christie, H. D'Agnes, K. Lowry and N. Milne, 2005. 'Designing ICM Projects for Sustainability: Lessons from the Philippines and Indonesia.' *Ocean & Coastal Management* 48: 271–296.

White, A.T., C.A. Courtney and A.M. Salamanca, 2002. 'Experience with Marine Protected Area Planning and Management in the Philippines.' *Coastal Management* 30: 1–26.

Wiegele, K.L., 2006. 'Catholics Rich in Spirit: El Shaddai's Modern Engagements.' *Philippine Studies* 54: 495–520.

Wildlife Extra, 2007. '200 Sea Turtles Found on Board Chinese Fishing Boat.' Viewed 1 October 2007 at http://www.wildlifeextra.com/go/news/turtle-poaching128.html

Wolf, E.R., 1999. *Envisioning Power: Ideologies of Dominance and Crisis*. Berkeley (CA): University of California Press.

Wolters, W., 1983. *Politics, Patronage and Class Conflict in Central Luzon*. The Hague: Institute of Social Studies.

WVA (World Vision Australia), 2006. 'Gutpela Tingting na Sindaun: Papua New Guinean Perspectives on a Good Life.' Victoria: WVA Advocacy and Public Influence Unit.

Wright, T., 1978. Querencia Calamian: A Geographic Approach to Integrated Regional Conservation and Development in Calamianes, Palawan, Philippines. Honolulu: University of Hawaii (Ph.D. thesis).

WWF (Worldwide Fund for Nature), 2008. 'Small Boats, Big Problems.' Gland: WWF International.

Zayas, C.N., 1994. 'Pangayaw and Tumandok in the Maritime World of the Visayan Islanders.' In I. Ushijima and C. Neri Zayas (eds), op. cit.

Index

www.ingramcontent.com/pod-product-compliance
Lightning Source LLC
Chambersburg PA
CBHW061244270326
41928CB00041B/3411